Inorganic
Structural Chemistry

Inorganic Chemistry
A Textbook Series

Editorial Board
Gerd Meyer, *University of Hannover*
Akira Nakamura, *Osaka University*

Advisory Board
Ginya Adachi, *Osaka University*
Ken Poeppelmeier, *Northwestern University*

Ulrich Müller
Curriculum vitae

Born in 1940 in Bogotá, Colombia. School attendance in Bogotá, then in Elizabeth, New Jersey, and finally in Fellbach, Germany. Studied chemistry at the Technische Hochschule in Stuttgart, Germany, obtaining the degree of Diplom-Chemiker in 1963. Work on the doctoral thesis in inorganic chemistry was performed in Stuttgart and at Purdue University in West Lafayette, Indiana, in the research groups of K. Dehnicke and K.S. Vorres, respectively. The doctor's degree in natural sciences (Dr. rer.nat.) was awarded by the Technische Hochschule Stuttgart in 1966. Subsequent post-doctoral work in crystallography and crystal structure determinations was performed in the research group of H. Bärnighausen at the University of Karlsruhe, Germany. Appointed in 1972 as professor of inorganic chemistry at the University of Marburg, Germany. Helped installing a graduate school of chemistry as visiting professor at the University of Costa Rica from 1975 to 1977. Courses in spectroscopic methods were repeatedly given at different universities in Costa Rica, Brazil and Chile. Main areas of scientific interest: synthetic inorganic chemistry, crystallography and crystal structure systematics, crystal structure determinations, infrared and Raman spectroscopy. Co-author of *Schwingungsspektroskopie*, a textbook about the application of vibrational spectroscopy, and of *Schwingungsfrequenzen I and II* (tables of characteristic molecular vibrational frequencies).

Inorganic Structural Chemistry

Ulrich Müller

University of Marburg, Germany

JOHN WILEY & SONS
Chichester · New York · Brisbane · Toronto · Singapore

German version
Anorganische Strukturchemie
First edition 1991
Second edition 1992
© B.G. Teubner Stuttgart 1991
The author was awarded the Prize for Chemical Literature for this book by the
Verband der Chemischen Industrie (German Federation of Chemical Industries) in 1992

Copyright © 1993 by John Wiley & Sons Ltd,
Baffins Lane, Chichester,
West Sussex PO19 1UD, England

Other Wiley Editorial Offices

John Wiley & Sons, Inc., 605 Third Avenue,
New York, NY 10158-0012, USA

Jacaranda Wiley Ltd, G.P.O. Box 859, Brisbane,
Queensland 4001, Australia

John Wiley & Sons (Canada) Ltd, 22 Worcester Road,
Rexdale, Ontario M9W 1L1, Canada

John Wiley & Sons (SEA) Pte Ltd, 37 Jalan Pemimpin #05-04,
Block B, Union Industrial Building, Singapore 2057

British Library Cataloguing in Publication Data

A catalogue record for this book is available from the British Library

ISBN 0 471 93379 1 (Cloth)
ISBN 0 471 93717 7 (Pbk)

Typeset with the programs TEX by D.E. Knuth and LATEX by L. Lamport.
Most of the figures were produced with the program ATOMS by E. Dowty

Printed and Bound in Great Britain by Bookcraft Bath Ltd, Midsomer Norton, Avon

Contents

Preface

Given the increasing quantity of knowledge in all areas of science, the imparting of this knowledge must necessarily concentrate on general principles and laws, while details must be restricted to important examples. A textbook should be reasonably small, but essential aspects of the subject may not be neglected, traditional foundations must be considered, and modern developments should be included. This introductory text is an attempt to present inorganic structural chemistry in this way. Compromises cannot be avoided; some sections may be shorter, while others may be longer than some experts in this area may deem appropriate.

Chemists predominantly think in illustrative models: they like to "see" structures and bonds. Modern bond theory has won its place in chemistry, and is given proper attention in chapter 9. However, with its extensive calculations it corresponds more to the way of thinking of physicists, and, furthermore, it is still often unsatisfactory when structural details are to be understood or predicted. For everyday use, simple models such as those treated in chapters 7, 8, and 12 are usually more useful to a chemist: "The peasant who wants to harvest in his lifetime cannot wait for the *ab initio* theory of weather. Chemists, like peasants, believe in rules, but cunningly manage to interpret them as occasion demands" (H.G. VON SCHNERING [102]).

This book is mainly addressed to advanced students of chemistry. Basic chemical knowledge concerning atomic structure, chemical bond theory and structural aspects is required. Parts of the text are based on a course on inorganic crystal chemistry by Prof. H. Bärnighausen at the University of Karlsruhe. I am grateful to him for permission to use the manuscript of his course, for numerous suggestions, and for his encouragement. For discussions and suggestions I also thank Prof. D. Babel, Prof. K. Dehnicke, Prof. C. Elschenbroich, Prof. D. Reinen and Prof. G. Weiser. I thank Ms. J. Gregory for reviewing and correcting the English version of the manuscript.

Ulrich Müller   Marburg, Germany, October 1992

1 Introduction

Structural chemistry or *stereochemistry* is the science of the structures of chemical compounds, the latter term being used mainly when the structures of molecules are concerned. Structural chemistry deals with the elucidation and description of the spatial order of atoms in a compound, with the explanation of the reasons that lead to this order, and with the properties resulting thereof. It also includes the systematic ordering of the recognized structure types and the disclosure of relationships among them.

Structural chemistry is an essential part of modern chemistry in theory and practice. To understand the processes taking place during a chemical reaction and to render it possible to design experiments for the synthesis of new compounds, a knowledge of the structures of the compounds involved is essential. Chemical and physical properties of a substance can only be understood when its structure is known. The enormous influence that the structure of a material has on its properties can be seen by the comparison of graphite and diamond: both consist only of carbon, and yet they differ widely in their physical and chemical properties.

The most important experimental task in structural chemistry is the *structure determination*. Structure determination is the analytical aspect of structural chemistry; the usual result is a static model. The elucidation of the spatial rearrangements of atoms during a chemical reaction is much less accessible experimentally. *Reaction mechanisms* deal with this aspect of structural chemistry in the chemistry of molecules. *Topotaxy* is concerned with chemical processes in solids, in which structural relations exist between the orientation of educts and products. Neither dynamic aspects of this kind nor the experimental methods of structure determination are subjects of this book.

Crystals are distinguished by the regular, periodic order of their components. In the following we will focus much attention on this order. However, this should not lead to the impression of a perfect order. Real crystals contain numerous faults, their number increasing with temperature. Atoms can be missing or misplaced, and dislocations and other imperfections can occur.

2 Description of Chemical Structures

In order to specify the structure of a chemical compound, we have to describe the spatial distribution of the atoms in an adequate manner. This can be done with the aid of chemical nomenclature, which is well developed, at least for small molecules. However, for solid state structures, there exists no systematic nomenclature which allows us to specify structural facts. One manages with the specification of *structure types* in the following manner: "magnesium fluoride crystallizes in the rutile type", which expresses for MgF_2 a distribution of Mg and F atoms corresponding to that of Ti and O atoms in rutile. Every structure type is designated by an arbitrarily chosen representative. How structural information can be expressed in formulas is treated in section 2.1.

Graphic representations are useful. One of these is the much used valence bond formula, which allows a pithy representation of essential structural aspects of a molecule. More exact and more illustrative are perspective, true-to-scale figures, in which the atoms are drawn as balls or — if the always present thermal vibrations are to be expressed — as ellipsoids. To achieve a better view, the balls or ellipsoids are plotted to a smaller scale than that corresponding to their effective sizes. Covalent bonds are represented as sticks. The size of a thermal ellipsoid is chosen to represent the probability of finding the atom averaged over time (usually 50 % probability of finding the center of the atom within the ellipsoid; cf. Fig. 1b). For more complicated structures the perspective image

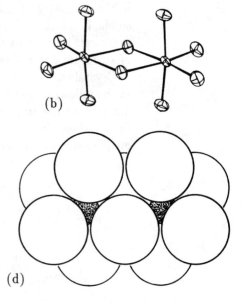

Fig. 1 Graphic representations for a molecule of $(UCl_5)_2$. (a) Valence bond formula. (b) Perspective view with ellipsoids of thermal motion. (c) Coordination polyhedra. (d) Emphasis of the space requirements of the chlorine atoms

can be made clearer with the aid of a stereoscopic view (cf. Fig. 18, p. 41). Different types of drawings can be used to stress different aspects of a structure (Fig. 1).

Quantitative specifications are made with numeric values for interatomic distances and angles. The interatomic distance is defined as the distance between the nuclei of two atoms in their mean positions (mean positions of the thermal vibration). The most common method to determine interatomic distances experimentally is X-ray diffraction from single crystals. Other methods include neutron diffraction from crystals and, for small molecules, electron diffraction and microwave spectroscopy with gaseous samples. X-ray diffraction determines not the positions of the atomic nuclei but the positions of the centers of the negative charges of the atomic electron shells, because X-rays are diffracted by the electrons of the atoms. However, the negative charge centers coincide almost exactly with the positions of the atomic nuclei, except for covalently bonded hydrogen atoms. To locate hydrogen atoms exactly, neutron diffraction is also more appropriate than X-ray diffraction for another reason: X-rays are diffracted by the large number of electrons of heavy atoms to a much larger extent, so that the position of H atoms in the presence of heavy atoms can be determined only with low reliability. This is not the case for neutrons, as they interact with the atomic nuclei. (Because neutrons suffer incoherent scattering from H atom nuclei to a larger extent than from D atom nuclei, neutron scattering is performed with deuterated compounds.)

2.1 Coordination Numbers and Coordination Polyhedra

The coordination number (c.n.) and the coordination polyhedron serve to characterize the immediate surroundings of an atom. The *coordination number* specifies the number of coordinated atoms; these are the closest neighboring atoms. For many compounds there are no difficulties in stating the coordination numbers for all atoms. However, it is not always clear up to what limit a neighboring atom is to be counted as a closest neighbor. For instance, in metallic antimony every Sb atom has three neighboring atoms at distances of 291 pm and three others at distances of 336 pm, which is only 15 % more. In this case it helps to specify the coordination number by 3+3, the first number referring to the number of neighboring atoms at the shorter distance.

Stating the coordination of an atom as a single number is not very informative in more complicated cases. However, specifications of the following kind can be made: in white tin an atom has four neighboring atoms at a distance of 302 pm, two at 318 pm and four at 377 pm. Several propositions have been made to calculate a mean or "effective" coordination number (e.c.n. or ECoN) by adding all surrounding atoms with a weighting scheme, in that the atoms are not counted as full atoms, but as fractional atoms with a number between 0 and 1; this number is closer to zero when the atom is further away. Frequently a gap can be found in the distribution of the interatomic distances of the neighboring atoms: if the shortest distance to a neighboring atom is set equal to 1, then often further atoms are found at distances between 1 and 1.3, and after them follows a gap in which no atoms are found. According to a proposition of G. BRUNNER and D. SCHWARZENBACH an atom at the distance of 1 obtains the weight 1, the first atom beyond the gap obtains zero weight, and all intermediate atoms are included with weights that are calculated from their distances by linear interpolation [70]: e.c.n. $= \sum_i (d_g - d_i)/(d_g - d_1)$; $d_1 =$ distance to the closest atom, $d_g =$ distance to the first atom beyond the gap, $d_i =$ distance to the i-th atom in the region between d_1 and d_g. For Example for antimony: taking $3 \times d_1 = 291$, $3 \times d_i = 336$ and $d_g = 391$ pm one obtains e.c.n. $=$ 4.65. The method is however of no help when no clear gap can be discerned. A mathematically unique method of calculation considers the *domain of influence* (also called *Wirkungsbereich*, VORONOI polyhedron, WIGNER-SEITZ cell, or DIRICHLET domain). The domain is constructed by connecting the atom in question with all surrounding atoms; the set of planes perpendicular to the connecting lines and passing through their midpoints forms the domain of influence, which is a convex polyhedron. In this way, a polyhedron face can be assigned to every neighboring atom, the area of the face serving as measure for the weighting [71,72,73]. Other formulas have also been derived, e.g. ECoN $= \sum_i \exp[1 - (d_i/d_1)^n]$, with $n = 5$ or 6, $d_i =$ distance to the i-th atom and $d_1 =$ shortest distance or $d_1 =$ fictive standard distance [69,74]. For example, using the last-mentioned method with the formula by R. HOPPE we obtain ECoN $=$ 6.5 for white tin and ECoN $= 4.7$ for antimony.

The kind of bond between neighboring atoms also has to be considered. For instance, the coordination number for a chlorine atom in the CCl_4 molecule is 1 when only the covalently bonded C atom is counted, but it is 4 (1 C + 3 Cl) when all atoms "in contact" are counted. In the case of molecules one will tend to count only covalently bonded atoms as coordinated atoms. In the case of crystals consisting of monoatomic ions usually only the anions immediately adjacent to a cation and the cations immediately adjacent to an anion are considered, even when there are contacts between anions and anions or between cations and cations. In this way, an I^- ion in LiI (NaCl type) is assigned the coordination number 6, whereas it is 18 when the 12 I^- ions with which it is also in contact are included. In case of doubt, one should always specify exactly what is to be included in the coordination sphere.

The *coordination polyhedron* results when the centers of mutually adjacent coordinated atoms are connected with one another. For every coordination number typical coordination polyhedra exist (Fig. 2). In some cases, several coordination polyhedra for a given coordination number differ only slightly, even though this may not be obvious at first glance; by minor displacements of atoms one polyhedron may be converted into another. For instance, a trigonal bipyramid can be converted into a tetragonal pyramid by displacements of four of the coordinated atoms (Fig. 22, p. 60).

Larger structural units can be described by connected polyhedra. Two poly-

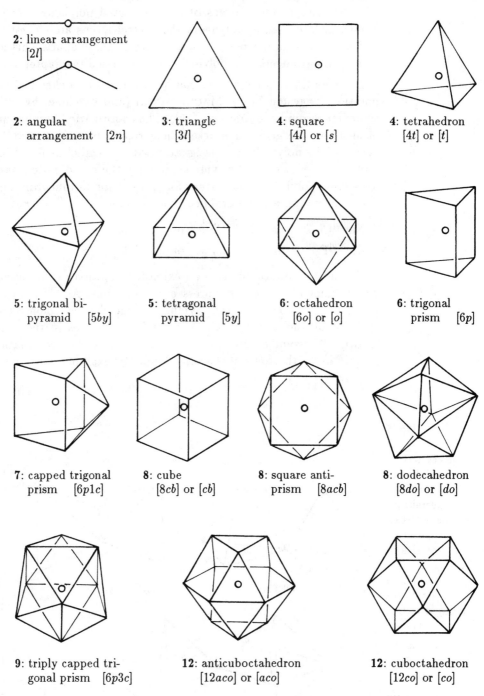

2: linear arrangement [2*l*]

2: angular arrangement   [2*n*]

3: triangle [3*l*]

4: square [4*l*] or [*s*]

4: tetrahedron [4*t*] or [*t*]

5: trigonal bi-pyramid   [5*by*]

5: tetragonal pyramid   [5*y*]

6: octahedron [6*o*] or [*o*]

6: trigonal prism   [6*p*]

7: capped trigonal prism   [6*p*1*c*]

8: cube [8*cb*] or [*cb*]

8: square anti-prism   [8*acb*]

8: dodecahedron [8*do*] or [*do*]

9: triply capped tri-gonal prism   [6*p*3*c*]

12: anticuboctahedron [12*aco*] or [*aco*]

12: cuboctahedron [12*co*] or [*co*]

Fig. 2

The most important coordination polyhedra and their symbols; for explanation of the symbols see p. 7

hedra can be joined by a common vertex, a common edge, or a common face
(Fig. 3). The common atoms of two connected polyhedra are called bridging
atoms. In face-sharing polyhedra the central atoms are closest to one another
and in vertex-sharing polyhedra they are furthest apart. Further details concerning the connection of polyhedra are discussed in chapter 15.

The coordination conditions can be expressed in a chemical formula using a
notation suggested by F. MACHATSCHKI (and extended by several other authors; for recommendations see [75]). The coordination number and polyhedron
of an atom are given in brackets in a right superscript next to the element
symbol. The polyhedron is designated with a symbol as listed in Fig. 2. Short
forms can be used for the symbols, namely the coordination number alone or,
for simple polyhedra, the letter alone, e.g. t for tetrahedron, and in this case
the brackets can also be dropped. For example:

$$Na^{[6o]}Cl^{[6o]} \quad \text{or} \quad Na^{[6]}Cl^{[6]} \quad \text{or} \quad Na^o Cl^o$$
$$Ca^{[8cb]}F_2^{[4t]} \quad \text{or} \quad Ca^{[8]}F_2^{[4]} \quad \text{or} \quad Ca^{cb}F_2^t$$

For more complicated cases an extended notation can be used, in which the
coordination of an atom is expressed in the manner $A^{[m,n;p]}$. For m, n, and p the
polyhedra symbols are taken. Symbols before the semicolon refer to polyhedra
spanned by the atoms B, C... , in the sequence as in the chemical formula
$A_a B_b C_c$. The symbol after the semicolon refers to the coordination of the atom
in question with atoms of the same kind. For example perovskite:

$$Ca^{[,12co]}Ti^{[,6o]}O_3^{[4l,2l;8p]} \qquad \text{(cf. Fig. 124, p. 198)}$$

(cf. Fig. 124, p. 198)

Fig. 3
Examples for
the connection
of polyhedra.
(a) Two
tetrahedra
sharing a vertex.
(b) Two
tetrahedra
sharing an edge.
(c) Two
octahedra
sharing a vertex.
(d) Two
octahedra
sharing a face.
For two
octahedra
sharing an edge
see Fig. 1

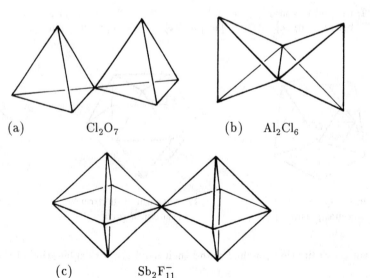

(a) Cl_2O_7 (b) Al_2Cl_6

(c) $Sb_2F_{11}^-$

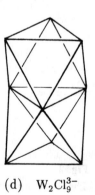

(d) $W_2Cl_9^{3-}$

Since Ca is not directly surrounded by Ti atoms, the first polyhedron symbol is dropped; however, the first comma cannot be dropped to keep clear that the 12*co* refers to a cuboctahedron formed by 12 O atoms. Ti is not directly surrounded by Ca, but by 6 O atoms forming an octahedron. O is surrounded in planar (square) coordination by four Ca, by two linearly arranged Ti and by eight O atoms forming a prism.

In addition to the polyhedra symbols listed in Fig. 2, further symbols can be constructed. The letters have the following meanings:

l	collinear or coplanar	*t*	tetrahedral	*do*	dodecahedral
		s	square	*co*	cuboctahedral
n	not collinear or coplanar	*o*	octahedral	*i*	icosahedral
		p	prismatic	*c*	capped
y	pyramidal	*cb*	cubic	*a*	anti-
by	bipyramidal	*FK*	Frank-Kasper polyhedron (Fig. 86)		

For example: $[3n]$ = three atoms not coplanar with the central atom as in NH_3; $[12p]$ = hexagonal prism. When lone electron pairs in polyhedra vertices are also counted, a symbolism in the following manner can be used: $[\psi - 4t]$ (same meaning as $[3n]$), $[\psi - 6o]$ (same as $[5y]$), $[2\psi - 6o]$ (same as $[4l]$).

When coordination polyhedra are connected to chains, layers or a three dimensional network, this can be expressed by the preceding symbols $\frac{1}{\infty}$, $\frac{2}{\infty}$ or $\frac{3}{\infty}$, respectively. Examples:

$$\frac{3}{\infty}Na^{[6]}Cl^{[6]} \qquad \frac{3}{\infty}Ti^{[o]}O_2^{[3l]} \qquad \frac{2}{\infty}C^{[3l]} \text{ (graphite)}$$

To state the existence of individual, finite atom groups, a $\frac{0}{\infty}$ can be set in front of the symbol. For their further specification, the following symbols may be used:

chain fragment	$\{f\}$ or $\wedge$
ring	$\{r\}$ or $\bigcirc$
cage	$\{k\}$ or $\varheartsuit$

For example: $Na_2\wedge S_3$; $\{k\}P_4$; $Na_3\bigcirc[P_3O_9]$.

Another type of notation, introduced by P. NIGGLI, uses fractional numbers in the chemical formula. The formula $TiO_{6/3}$ for instance means that every titanium atom is surrounded by 6 O atoms, each of which is coordinated to 3 Ti atoms. Another example is: $NbOCl_3 = NbO_{2/2}Cl_{2/2}Cl_{2/1}$ which has coordination number 6 for the niobium atom (= 2 + 2 + 2 = sum of the numerators), coordination number 2 for the O atom and coordination numbers 2 and 1 for the two different kinds of Cl atoms (cf. Fig. 97, p. 168).

2.2 The Description of Crystal Structures

In a crystal atoms are joined to form a larger network in a periodically ordered way. The spatial order of the atoms is called the *crystal structure*. When we connect the periodically repeated atoms of one kind in three space directions to a three-dimensional grid, we obtain the *crystal lattice*. The crystal lattice

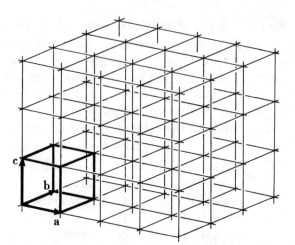

Fig. 4
Primitive cubic
crystal lattice.
One unit cell is
outlined

represents a three-dimensional order of points; all points of the lattice are completely equivalent and have identical surroundings. We can think of the crystal lattice as generated by periodically repeating a small parallelepiped in three dimensions without gaps (Fig. 4; parallelepiped = body limited by six faces that are parallel in pairs). The parallelepiped is called the *unit cell.*

The unit cell can be defined by three *base vectors* labeled **a**, **b** and **c**. The crystal lattice corresponds to the complete set of all linear combinations $\mathbf{r} = u\mathbf{a}+v\mathbf{b}+w\mathbf{c}$, u, v, w comprising all integer numbers. The lengths a, b, and c of the base vectors and the angles α, β, and γ between them are the *lattice parameters* (or lattice constants). There is no unique way to choose the unit cell for a given crystal structure, as is illustrated for a two dimensional example in Fig. 5. To achieve a standardization in the description of crystal structures, certain conventions for the selection of the unit cell have been settled upon in crystallography:

1. The unit cell is to show the symmetry of the crystal, i.e. the base vectors are to be chosen parallel to symmetry axes or perpendicular to symmetry planes.

2. For the origin of the unit cell a geometrically unique point is selected, with priority given to a symmetry center.

Fig. 5
Periodical, two
dimensional arrangement
of A and X atoms. The
whole pattern can be
generated by repeating
any one of the plotted
unit cells.

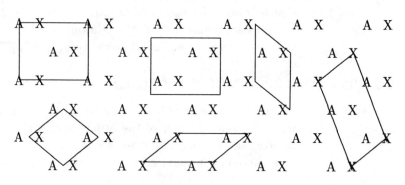

3. The base vectors should be as short as possible. This also means: the cell volume should be as small as possible, and the angles between them should be as close as possible to 90°.

4. As far as the angles between the base vectors deviate from 90°, they are either chosen to be all larger or all smaller than 90° (preferably > 90°).

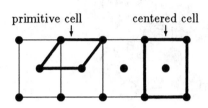

primitive cell   centered cell

A unit cell having the smallest possible volume is called a *primitive* cell. For reasons of symmetry according to rule 1 and contrary to rule 3, a primitive cell is not always chosen, but instead a *centered* cell, which is *double, triple* or *fourfold primitive*, i.e. its volume is larger by a corresponding factor. The centered cells to be considered are shown in Fig. 6.

Aside from the conventions mentioned for the cell choice, further rules have been developed to achieve standardized descriptions of crystal structures [76]. They should be followed to assure a systematic and comparable registration of the data and to allow for the inclusion in data bases. Nevertheless, sometimes there are objective reasons to disregard the standards, e.g. when the relationships between different structures are to be pointed out.

Specification of the lattice parameters and the positions of all atoms contained in the unit cell is sufficient to characterize all essential aspects of a crystal structure. A unit cell can only contain an integral number of atoms. When stating the contents of the cell one refers to the chemical formula, i.e. the number of "formula units" per unit cell is given; this number is usually termed Z. How the atoms are to be counted is shown in Fig. 7. As a matter of fact, a cell containing a non-integer number of atoms is sometimes stated, namely in the case of "nonstoichiometric" compounds. In this kind of compounds certain atoms are statistically present or not in different cells, so that on average a non-integer number of atoms per cell results.

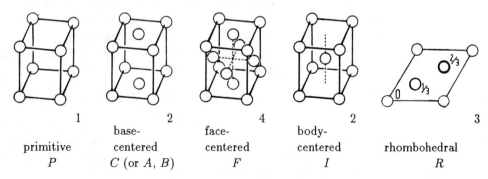

1	2	4	2	3
primitive	base-centered	face-centered	body-centered	rhombohedral
P	C (or A, B)	F	I	R

Fig. 6
Centered unit cells and their symbols. The numbers specify how many fold primitive the respective cell is. The numbers in the R cell refer to the height in the cell

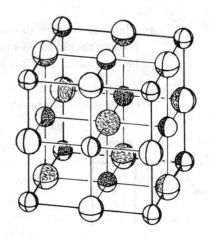

Fig. 7
The way to count the contents of a unit cell
for the example of the face-centered unit cell of
NaCl: 8 Na^+ ions in 8 vertices, each of which belongs to 8 adjacent cells makes $8/8 = 1$; 6 Na^+
ions in the centers of 6 faces belonging to two adjacent cells each makes $6/2 = 3$. 12 Cl^- ions in
the centers of 12 edges belonging to 4 cells each
makes $12/4 = 3$; 1 Cl^- ion in the cube center,
belonging only to this cell. Total: 4 Na^+ and 4
Cl^- ions or four formula units of NaCl ($Z = 4$).

2.3   Atomic Coordinates

The position of an atom in the unit cell is specified by a set of *atomic coordinates*,
i.e. by three coordinates x, y and z. These refer to a coordinate system that is
defined by the base vectors of the unit cell. The unit length taken along every
one of the coordinate axes corresponds to the length of the respective base
vector. The coordinates x, y and z for every atom within the unit cell thus
have values between 0.0 and 1.0. The coordinate system is *not* a cartesian one;
the coordinate axes can be inclined to one another and the unit lengths on
the axes may differ from each other. Addition or substraction of an *integral*
number to a coordinate value generates the coordinates of an equivalent atom
in a different unit cell. For instance, the coordinate triplet $x = 1.27$, $y = 0.52$
and $z = -0.10$ specifies the position of an atom in a cell neighboring the origin
cell, namely in the direction $+\mathbf{a}$ and $-\mathbf{c}$; this atom is equivalent to the atom at
$x = 0.27$, $y = 0.52$ and $z = 0.90$ in the origin cell.

Commonly, only the atomic coordinates for the atoms in one *asymmetric unit*
are listed. Atoms that can be "generated" from these by existing symmetry
elements are not listed. Which symmetry elements are present is revealed by
stating the *space group* (cf. section 18.3). When the lattice parameters, the
space group, and the atomic coordinates are known, all structural details can
be deduced. In particular, all interatomic distances and angles can be calculated.

The following formula can be used to calculate the distance d between two
atoms from the lattice parameters and atomic coordinates:

$$d = \sqrt{(a\Delta x)^2 + (b\Delta y)^2 + (c\Delta z)^2 + 2bc\Delta y\Delta z \cos\alpha + 2ac\Delta x\Delta z \cos\beta + 2ab\Delta x\Delta y \cos\gamma}$$

$\Delta x = x_2 - x_1$, $\Delta y = y_2 - y_1$ and $\Delta z = z_2 - z_1$ are the differences between the
coordinates of the two atoms. The angle ω at atom 2 in a group of three atoms
1, 2 and 3 can be calculated from the three distances d_{12}, d_{23} and d_{13} between
them according to the cosine formula:

$$\cos\omega = -\sqrt{\frac{d_{13}^2 - d_{12}^2 - d_{23}^2}{2d_{12}d_{23}}}$$

When specifying atomic coordinates, interatomic distances etc., the corresponding standard deviations should also be given, which serve to express the reliability of their experimental determination. The commonly used notation, such as "$d = 235.1(4)$ pm" states a standard deviation of 4 units for the last digit, i.e. the standard deviation in this case amounts to 0.4 pm. Standard deviation is a term in statistics. When a standard deviation σ is linked to some value, the probability of the true value being within the limits $\pm\sigma$ of the stated value is 68.3 %. The probability of being within $\pm 2\sigma$ is 95.4 %, and within $\pm 3\sigma$ is 99.7 %.

2.4 Isotypism

The crystal structures of two compounds are *isotypic* if their atoms are distributed in a like manner and if they have the same symmetry. One of them can be generated from the other one if atoms of an element are substituted by atoms of another element without changing their positions in the crystal structure. The absolute values of the lattice dimensions and the interatomic distances may differ, and *small* variations are permitted for the atomic coordinates. The angles between the crystallographic axes and the relative lattice dimensions (axes ratios) must be similar. Two isotypic structures exhibit a one-to-one relation for all atomic positions and have coincident geometric conditions. If, in addition, the chemical bonding conditions are also similar, then the structures also are *crystal chemical isotypic*.

The ability of two compounds which have isotypic structures to form mixed crystals, i.e. when the exchange process of the atoms can actually be performed continuously, has been termed *isomorphism*. However, because this term also is used for some other phenomena, it has been recommended that its use be discontinued in this context.

Two structures are *homeotypic* if they are similar, but fail to fulfil the aforementioned conditions for isotypism because of different symmetry, because corresponding atomic positions are occupied by several different kinds of atoms (substitution derivatives) or because the geometric conditions differ (different axes ratios, angles, or atomic coordinates). An example of substitution derivatives is: C (diamond)–ZnS (sphalerite)–Cu_3SbS_4 (famatinite).

If two ionic compounds have the same structure type, but in such a way that the cationic positions of one compound are taken by the anions of the other and vice versa ("exchange of cations and anions"), then they sometimes are called "antitypes". For example: in Li_2O the Li^+ ions occupy the same positions as the F^- ions in CaF_2, while the O^{2-} ions take the same positions as the Ca^{2+} ions; Li_2O crystallizes in the "anti-CaF_2 type".

2.5   Problems

2.1 Calculate effective coordination numbers (e.c.n.) with the formula given on page 4 for:
(a) Tellurium, $4 \times d_1 = 283$ pm, $2 \times d_2 = 349$ pm, $d_g = 444$ pm;
(b) Gallium, $1 \times d_1 = 247$ pm, $2 \times d_2 = 270$ pm, $2 \times d_3 = 274$ pm, $2 \times d_4 = 279$ pm, $d_g = 398$ pm;
(c) Tungsten, $8 \times d_1 = 274.1$ pm, $6 \times d_2 = 316.5$ pm, $d_g = 447.6$ pm.

2.2 Include specifications of the coordination of the atoms in the following formulas:
(a) $FeTiO_3$, Fe and Ti octahedral, O being coordinated by 2 Fe and by 2 Ti in a nonlinear arrangement in each case;
(b) $CdCl_2$, Cd octahedral, Cl trigonal-nonplanar;
(c) MoS_2, Mo trigonal-prismatic, S trigonal-nonplanar;
(d) Cu_2O, Cu linear, O tetrahedral;
(e) PtS, Pt square, S tetrahedral;
(f) $MgCu_2$, Mg FRANK-KASPER polyhedron with c.n. 16, Cu icosahedral;
(g) $Al_2Mg_3Si_3O_{12}$, Al octahedral, Mg dodecahedral, Si tetrahedral;
(h) UCl_3, U tricapped trigonal-prismatic, Cl 3-nonplanar.

2.3 Give the symbols stating the kind of centering of the unit cells of CsCl, NaCl and sphalerite (Fig. 14), CaF_2 and rutile (Fig. 17), CaC_2 and NaN_3 (Fig. 19, heavily outlined cells), ReO_3 (Fig. 91).

2.4 Give the number of formula units per unit cell for:
CsCl (Fig. 14), ZnS (Fig. 14), TiO_2 (rutile, Fig. 17), $ThSi_2$ (Fig. 63), ReO_3 (Fig. 91), α-$ZnCl_2$ (Fig. 128).

2.5 What is the I–I bond length in solid iodine? Unit cell parameters: $a = 714$, $b = 469$, $c = 978$ pm, $\alpha = \beta = \gamma = 90°$. Atomic coordinates: $x = 0.0$, $y = 0.1543$, $z = 0.1174$; A symmetrically equivalent atom is at $-x, -y, -z$.

2.6 Calculate the bond lengths and the bond angle at the central atom of the I_3^- ion in RbI_3. Unit cell parameters: $a = 1091$, $b = 1060$, $c = 665.5$ pm, $\alpha = \beta = \gamma = 90°$. Atomic coordinates: I(1), $x = 0.1581$, $y = 0.25$, $z = 0.3509$; I(2), $x = 0.3772$, $y = 0.25$, $z = 0.5461$; I(3), $x = 0.5753$, $y = 0.25$, $z = 0.7348$.

In the following problems the positions of symmetrically equivalent atoms (due to space group symmetry) may have to be considered; they are given as coordinate triplets to be calculated from the generating position x, y, z. To obtain positions of adjacent (bonded) atoms, some atomic positions may have to be shifted to a neighboring unit cell.

2.7 MnF_2 crystallizes in the rutile type with $a = b = 487.3$ pm and $c = 331.0$ pm. Atomic coordinates: Mn at $x = y = z = 0$; F at $x = y = 0.3050$, $z = 0.0$. Symmetrically equivalent positions: $-x, -x, 0$; $0.5-x, 0.5+x, 0.5$; $0.5+x, 0.5-x, 0.5$. Calculate the two different Mn–F bond lengths (< 250 pm) and the F–Mn–F bond angle referring to two F atoms having the same x and y coordinates and z differing by 1.0.

2.8 $WOBr_4$ is tetragonal, $a = b = 900.2$ pm, $c = 393.5$ pm, $\alpha = \beta = \gamma = 90°$. Calculate the W–Br, W=O and W$\cdots$O bond lengths and the O=W–Br bond angle. Make a true-to-scale drawing (1 or 2 cm per 100 pm) of projections on to the ab and the ac plane, including atoms up to a distance of 300 pm from the z axis and covering $z = -0.5$ to $z = 1.6$. Draw atoms as circles and bonds as heavy lines (bonds being atomic contacts shorter than 300 pm). What is the coordination polyhedron of the W atom?
Symmetrically equivalent positions: $-x, -y, z$; $-y, x, z$; $y, -x, z$; atomic coordinates:

	x	y	z
W	0.0	0.0	0.0779
O	0.0	0.0	0.529
Br	0.2603	0.0693	0.0

2.9 Calculate the Zr–O bond lengths in baddeleyite (ZrO_2), considering only inter-atomic distances shorter than 300 pm. What is the coordination number of Zr?

Lattice parameters: $a = 514.5$, $b = 520.7$, $c = 531.1$ pm, $\beta = 99.23°$, $\alpha = \gamma = 90°$; symmetrically equivalent positions: $-x, -y, -z$; $x, 0.5-y, 0.5+z$; $-x, 0.5+y, 0.5-z$; atomic coordinates:

	x	y	z
Zr	0.2758	0.0411	0.2082
O(1)	0.0703	0.3359	0.3406
O(2)	0.5577	0.2549	0.0211

3 Polymorphism and Phase Diagrams

3.1 Polymorphism

Molecules having the same composition but different structures are called iso-mers. The corresponding phenomenon for crystalline solids is called *polymor-phism*; the different structures are the *modifications* or *polymorphic forms*. Mod-ifications differ not only in the spatial arrangement of their atoms, but also in their physical and chemical properties. The structural differences may comprise anything from minor variations in the orientations of molecules up to a com-pletely different atomic arrangement. Different modifications of a compound are frequently designated by lower case Greek letters $\alpha, \beta, \ldots$, e.g. α-sulfur, β-sulfur; polymorphic forms of minerals have in many cases been given triv-ial names, like α-quartz, β-quartz, tridymite, cristobalite, coesite, keatite, and stishovite for the SiO_2 forms.

Polymorphic forms with structures having different stacking sequences of like layers are called *polytypes*.

Which polymorphic form of a compound is formed depends on the prepara-tion and crystallization conditions: method of synthesis, temperature, pressure, kind of solvent, cooling or heating rate, crystallization from solution, fusion or gas phase, and presence of seed crystals are some of the factors of influence.

In many cases one modification can be converted to another by a change in temperature or pressure. In a similar way to the melting process, such a *phase transition* can involve a *sudden* change of the structure, volume, entropy and other thermodynamic functions at exactly defined temperature–pressure conditions; in this case it is a *first order* phase transition. It is accompanied by the exchange of conversion enthalpy with the surroundings. First order phase transitions exhibit *hysteresis*, i.e. the transition can take place some time after the temperature or pressure change giving rise to it. How fast the transfomation proceeds also depends on the formation or presence of sites of nucleation. The phase transition can proceed at an extremely slow rate. For this reason many thermodynamically unstable modifications are well-known and can be studied in conditions under which they should already have been transformed.

Second order phase transitions can extend over some temperature range. They take place with a continuous change in structure and show no hysteresis.

Enantiotropic phase transitions are reversible. If a modification is unstable at every temperature and every pressure, then its conversion into another modifi-cation is irreversible; such phase transitions are *monotropic*. When a compound that can form several modifications crystallizes, first a modification may form that is thermodynamically unstable under the given conditions; afterwards it converts to the more stable form (OSTWALD step rule). Selenium is an example: when elemental selenium forms by a chemical reaction in solution, it precipitates in a red modification that consists of Se_8 molecules; this then converts slowly into the stable, gray form that consists of polymer chain molecules. Potassium nitrate is another example: at room temperature β-KNO_3 is stable, but above 128 °C α-KNO_3 is stable. From an aqueous solution at room temperature α-KNO_3 crystallizes first, then, after a short while or when triggered by the slightest mechanical stress, it transforms to β-KNO_3.

The nucleation energy governs which modification crystallizes first. This energy depends on the surface energy. As a rule, nucleation energy decreases with decreasing surface energy. The modification having the smallest nucleation energy crystallizes first. As the surface energy depends sensitively on the adsorption of extraneous particles, the sequence of crystallization of polymorphic forms can be influenced by the presence of foreign matter.

3.2 Phase Diagrams

For phases that are in equilibrium with one another, the GIBBS' phase law holds:

$$F + P = C + 2$$

F is the number of degrees of freedom, i.e. the number of variables of state such as temperature and pressure that can be varied independently, P is the number of phases and C is the number of components. Components are to be understood as independent, pure substances (elements or compounds), from which the other compounds that eventually occur in the system can be formed. For example:

1. For pure water (one component, $C = 1$) $F + P = 3$ holds. When three phases are simultaneously in equilibrium with each other, e.g. vapor, liquid and ice, or vapor and two different modifications of ice, then $F = 0$; there is no degree of freedom, the three phases can coexist only at one fixed pressure and one fixed temperature ("triple point").

2. In the system iron/oxygen ($C = 2$), when two phases are present, e.g. Fe_3O_4 and oxygen, pressure and temperature can be varied ($F = 2$). When three phases are in equilibrium, e.g. Fe, Fe_2O_3 and Fe_3O_4, only one degree of freedom exists, and only the pressure *or* the temperature can be chosen freely.

A phase diagram in which pressure is plotted vs. temperature shows the existence ranges for the different phases of a system comprising only one component. Fig. 8 displays the phase diagram for water, in which the ranges of existence for vapor, liquid water and ten different modifications of ice are discernible, the latter being designated by Roman numerals. Within each of the marked fields only the corresponding phase is stable, but pressure and temperature can be varied independently (2 degrees of freedom). Along the delimiting lines two phases can

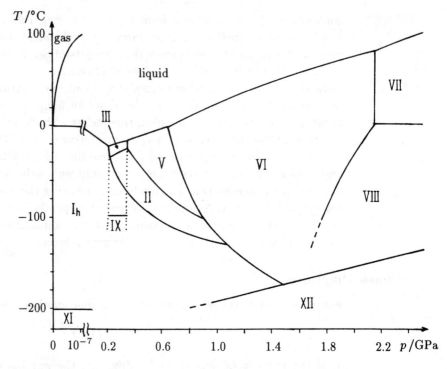

Fig. 8
Phase diagram for
H_2O. Dashed lines
are extrapolated;
they cannot be
measured because
of extremely slow
transition rates.
Additional meta-
stable phases are ice I_c in the low temperature section of ice I_h and ice IV within the region of ice V. The
line delimiting the gaseous/liquid sections continues up to the critical point at 374 °C and 22.1 MPa. The
liquid/ice VII line continues up to at least 20 GPa, where the melting point is 440 °C

coexist, and either the pressure or the temperature can be varied, whereas the
other one has to adopt the value specified by the diagram (one degree of freedom). At triple points there is no degree of freedom, pressure and temperature
have fixed values, but three phases are simultaneously in equilibrium with each
other.

In phase diagrams for two-component systems the composition is plotted vs.
one of the variables of state (pressure or temperature), the other one having a
constant value. Most common are plots of the composition vs. temperature at
normal pressure. Such phase diagrams differ depending on whether the components form solid solutions with each other or not or whether they combine to
form compounds.

The phase diagram for the system antimony/bismuth, in which mixed crystals (solid solutions) are formed, is shown in Fig. 9. Crystalline antimony and
bismuth are isotypic, and Sb and Bi atoms can occupy the atomic positions
in any proportion. The upper part of the diagram corresponds to the range of
existence of the liquid phase, i.e. a liquid solution of antimony and bismuth.
The lower part corresponds to the range of existence of the mixed crystals. In
between is a range in which liquid and solid coexist. On the upper side it is

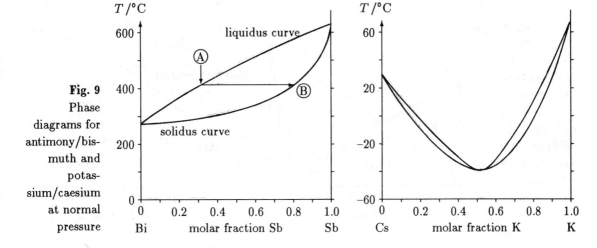

Fig. 9
Phase
diagrams for
antimony/bis-
muth and
potas-
sium/caesium
at normal
pressure

delimited by the *liquidus curve*, and on the lower side by the *solidus curve*. At
a given temperature, the liquid and the solid that are in equilibrium with one
another have different compositions. The compositions can be read from the
cross-points of the horizontal straight line marking the temperature in question with the *solidus* and the *liquidus* curves. Upon cooling an Sb/Bi melt with
a composition corresponding to the point marked A in Fig. 9, crystallization
begins when the temperature marked by the horizontal arrow is reached. The
composition of the mixed crystals that form is that of point B—the mixed
crystals have a higher Sb content than the melt.

The phase diagram potassium/caesium is an example of a system involving
the formation of mixed crystals with a temperature minimum (Fig. 9). The
right and left halves of the diagram are of the same type as the diagram for
antimony/bismuth. The minimum corresponds to a special point for which the
compositions of the solid and the liquid are the same. Other systems can have
the special point at a temperature maximum.

Limited formation of mixed crystals occurs when the two components have
different structures, as for example in the case of indium and cadmium. Mixed
crystals containing much indium and little cadmium have the structure of indium, while those containing little indium and much cadmium have the cadmium structure. At intermediate compositions a gap is observed, i.e. there are
no homogeneous mixed crystals, but instead a mixture of crystals rich in indium and crystals rich in cadmium is formed. This situation can even occur
when both of the pure components have the same structure, but mixed crystals
may not have any arbitrary composition. Copper and silver offer an example;
the corresponding phase diagram is shown in Fig. 10.

The phase diagram for aluminum/silicon (Fig. 10) is a typical example of a
system of two components that form neither solid solutions (except for very low
concentrations) nor a compound with one another, but are miscible in the liquid
state. As a special feature an acute minimum is observed in the diagram, the

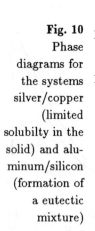

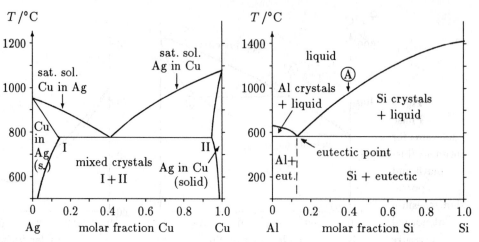

Fig. 10
Phase
diagrams for
the systems
silver/copper
(limited
solubilty in the
solid) and alu-
minum/silicon
(formation of
a eutectic
mixture)

eutectic point. It marks the melting point of the *eutectic mixture*, which is the mixture which has a lower melting point than either of the pure components or any other mixture. The *eutectic line* is the horizontal line that passes through the eutectic point. The area underneath is a region in which both components coexist as solids, i.e. in two phases.

A liquid solution of aluminum and silicon containing a molar fraction of 40 % aluminum and having a temperature of 1100 °C corresponds to point A in Fig. 10. Upon cooling the liquid we move downwards in the diagram (as marked by the arrow). At the moment we reach the liquidus line, pure silicon begins to crystallize. As a consequence, the composition of the liquid changes as it now contains an increasing fraction of aluminum; this corresponds to a leftward movement in the diagram. There the crystallization temperature for silicon is lower. According to the amount of crystallizing silicon, the temperature for the crystallization of further silicon decreases more and more until finally the eutectic point is reached, where both aluminum and silicon solidify. The term *incongruent solidification* serves to express the continuous change of the solidification temperature.

When the two components form compounds with each other, more complicated conditions arise. Fig. 11 shows the phase diagram for a system in which the two components, magnesium and calcium, form a compound, $CaMg_2$. At the composition $CaMg_2$ we observe a maximum which marks the melting point of this compound. On the left there is a eutectic point, formed by the components Mg and $CaMg_2$. On the right there is another eutectic point with the components Ca and $CaMg_2$. Both the left and right parts of the phase diagram in Fig. 11 correspond to the phase diagram of a simple eutectic system as in Fig. 10.

The system H_2O/HF exhibits an even more complicated phase diagram, as three compounds occur: $H_2O \cdot HF$, $H_2O \cdot 2\,HF$ and $H_2O \cdot 4\,HF$. In these compounds the particles H_3O^+, HF and F^- are joined with each other in different ways via hydrogen bridges. For two of the compounds we find maxima in the

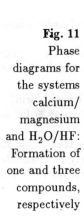

Fig. 11
Phase
diagrams for
the systems
calcium/
magnesium
and H$_2$O/HF:
Formation of
one and three
compounds,
respectively

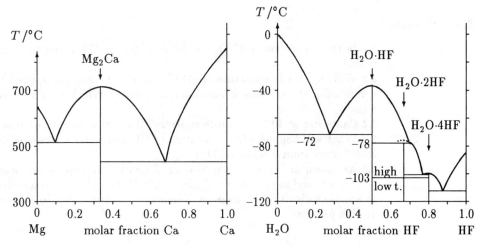

phase diagram (Fig. 11), in a similar way as for the compound CaMg$_2$. But there is no maximum at the composition H$_2$O·2 HF and no eutectic point between H$_2$O·2 HF and H$_2$O·HF; we can only discern a kink in the liquidus line. The expected maximum is "covered" (dashed line). The horizontal line running through the kink is called the *peritectic line*. The solids of composition H$_2$O·HF and H$_2$O·4 HF show congruent melting, i.e. they have definite melting temperatures according to the maxima in the diagram. The solid with the composition H$_2$O·2 HF, however, shows incongruent melting: at $-78\,°$C it decomposes to solid H$_2$O·HF and a liquid with a higher HF content. In addition, the compound H$_2$O·2 HF experiences a phase transition at $-103\,°$C from a "high temperature" to a low temperature modification; in the diagram this is expressed by the horizontal line at this temperature.

Phase diagrams give valuable information about the compounds that can form in a system of components. These compounds can then be prepared and studied. For the experimental determination of phase diagrams the following methods are used. In *differential thermo-analysis* (DTA) a sample of a given composition is heated or cooled slowly together with a thermally indifferent reference substance, and the temperatures of both substances are monitored continuously. When a phase transition occurs in the sample, the enthalpy of conversion is freed or absorbed and therefore a temperature difference shows up between the sample and the reference, thus indicating the phase transition. In *X-ray phase studies* an X-ray diffraction pattern is recorded continuously while the sample is being cooled or heated; when a phase transition occurs, changes in the diffraction pattern show up.

3.3   Problems

3.1 Is the conversion α-KNO_3 $\rightarrow$ β-KNO_3 (cf. p. 15) a first or second order phase transition?

3.2 Will ice at a temperature of $-10\,^\circ C$ melt if pressure is applied to it? If so, will it refreeze if the pressure is increased even more? Which modification would have to form?

3.3 Can water at $40\,^\circ C$ be made to freeze? If so, what modification of ice will form?

3.4 What will happen when a solution of HF and water containing 40 mole-% HF is cooled down from $0\,^\circ C$ to $-100\,^\circ C$?

3.5 As shown in Fig. 61, upon heating, β-quartz will experience phase transitions to β-tridymite and then to β-cristobalite at $870\,^\circ C$ and $1470\,^\circ C$, respectively. Is it feasible to achieve a direct interconversion β-quartz $\rightleftharpoons$ β-cristobalite by temperature variation at high pressure?

4 Structure, Energy and Chemical Bonding

4.1 Thermodynamic Stability

When the free enthalpy of reaction ΔG for the transformation of the structure of a compound to any other structure is positive, then this structure is *thermodynamically stable*. Since ΔG depends on the transition enthalpy ΔH and the transition entropy ΔS, and ΔH and ΔS in turn depend on pressure and temperature, a structure can be stable only within a certain range of pressures and temperatures. By variation of the pressure and/or the temperature, ΔG will eventually become negative relative to some other structure and a phase transition will occur. This may be a phase transition from a solid to another solid modification, or it may be a transition to another aggregate state.

According to the thermodynamic relations

$$\Delta G = \Delta H - T\Delta S \quad \text{and} \quad \Delta H = \Delta U + p\Delta V$$

the following rules can be given for the temperature and pressure dependence of thermodynamically stable structures:

1. With increasing temperature T structures with a low degree of order will be favored. Their formation involves a positive transition entropy ΔS and the value of ΔG then depends mainly on the term $T\Delta S$. For instance, among hexahalides such as MoF_6 two modifications are known in the solid state, one having molecules with well-defined orientations and the other having molecules rotating about their centers of gravity within the crystal. Since the order is lower for the latter modification, it is the thermodynamically stable one at higher temperatures. In the liquid state, the order is even lower and it is the lowest in the gaseous state. Raising the temperature thus will lead to melting and finally to evaporation of the substance.

2. Higher pressures p favor structures that occupy a lower volume, i.e. that have a higher density. As their formation involves a decrease of the volume (negative ΔV), ΔH will attain a negative value. For instance, diamond (density 3.51 g cm^{-3}) is more stable than graphite (density 2.26 g cm^{-3}) at very high pressures.

4.2 Kinetic Stability

A thermodynamically unstable structure can exist when its conversion to some other structure proceeds at a negligible rate. In this case we call the structure *metastable, inert* or *kinetically stable*. Since the rate constant k depends on the activation energy E_a and the temperature according to the ARRHENIUS equation,

$$k = k_0\, e^{-E_a/RT}$$

we have kinetic stability whenever a negligibly low k results from a large ratio E_a/RT. At sufficiently low temperatures any structure can be stabilized kinetically. Kinetic stability is not a well-defined term because the limit below which a conversion rate is to be considered negligible is arbitrary.

Glasses typically are metastable substances. Like crystalline solids they exhibit macroscopic form stability, but because of their structures and some of their physical properties they must be considered as liquids with a very high viscosity. Their transition to a thermodynamically more stable structure can only be achieved by extensive atomic movements, but atom mobility is severely hindered by cross-linking.

The structures and properties of numerous substances that are thermodynamically unstable under normal conditions are only known because they are metastable and therefore can be studied under normal conditions.

4.3   Chemical Bonding and Structure

Which spatial arrangement of atoms results in a stable or metastable structure depends decisively on the distribution of their electrons.

For noble gases the electronic configuration of a single atom is thermodynamically stable at normal conditions. Merely by packing the atoms closer together to form a liquid or a solid can a small amount of VAN DER WAALS energy still be released. The condensation and crystallization enthalpies being rather small, the magnitude of ΔG is governed already at relatively low temperatures by the term $T\Delta S$, and correspondingly the melting and boiling points are low.

For all other elements the electronic configuration of a single atom does not correspond to a thermodynamically stable state at normal conditions. Only at very high temperatures do single atoms occur in the vapor phase. At more ordinary temperatures atoms have to be linked to produce stable structures.

The electrons in an aggregate of atoms can only exist in certain definite energy states, just as in a single atom. These states are expressed mathematically by the eigenvalues of wave functions ψ. The wave functions result theoretically as solutions of the SCHRÖDINGER equation for the *complete set of all constituent atoms*. Although the exact mathematical solution of this equation poses insurmountable difficulties, we do have a well-founded knowledge about wave functions and thus about electrons in atomic systems. The knowledge is based on good mathematical approximations and on experimental data; we will discuss this more broadly in chapter 9. To begin with, we will restrict ourselves to the simplified scheme of two extreme kinds of chemical bonding, namely ionic and covalent bonds. However, we will also allow for intermediate states between these two extreme cases and we will consider the coexistence of both bonding types. As far as is relevant, we will also take into account the weaker ion–dipole, dipole–dipole and dispersion interactions.

The (localized) covalent bond is distinguished by its short range of action, which usually extends only from one atom to the next. However, within this range it is a strong bond. A near order arises around an atom; it depends on one hand on the tight interatomic bonds and on the other on the mutual repulsion of the valence electrons and on the space requirements of the bonded atoms.* When atoms are linked to form larger structures, the near order can result in a long range order in a similar way as the near order around a brick propagates to a long range order in a brick wall.

In a nonpolymer molecule or molecular ion a limited number of atoms is linked by covalent bonds. The covalent forces within the molecule are considerably stronger than all forces acting outwards. For this reason, when molecular structures are being considered, one commits only a small error when one acts as if the molecule would occur by itself and had no surroundings. Common experience as well as more detailed studies by KITAIGORODSKY [36] show that bond lengths and angles in molecules usually only undergo marginal alterations when the molecules assemble to a crystal. Only conformation angles are influenced more significantly in certain cases. Many properties of a molecular compound can therefore be explained from the molecular structure.

This is not equally valid for macromolecular compounds, in which a molecule consists of a nearly unlimited number of atoms. The interactions with surrounding molecules cannot be neglected in this case. For instance, for a substance consisting of thread-like macromolecules, it makes a difference to the physical properties whether the molecules are ordered in a crystalline manner or whether they are tangled.

Crystalline macromolecular substances can be classified according to the kind of connectivity of the covalently linked atoms as chain structures, layer structures and (three-dimensional) network structures. The chains or layers may be electrically uncharged molecules that interact with each other only by VAN DER WAALS forces, or they can be polyanions or polycations held together by intervening counter-ions. Network structures can also be charged, the counter-ions occupying cavities in the network. The structure of the chain, layer or network depends to a large extent on the covalent bonds and the resulting near order around each atom.

On the other hand, the crystal structures of ionic compounds with small molecular ions depend mainly on how space can be filled most efficiently by the ions, following the principle of cations around anions and anions around

---

*Frequently, *directionality* is a property attributed to the covalent bond which supposedly is taken to be the *cause* of the resulting structures. However, as the success of the valence electron pair repulsion theory shows, there exists no need to assume any orbitals directed a priori. The concept of directed orbitals is based on calculations in which hybridization is used as a *mathematical* aid. The popular use of hybridization models occasionally has created the false impression that hybridization is some kind of a process occurring prior to bond formation and commiting stereochemistry. See the discussion in section 9.2.

cations. Geometric factors such as the relative size of the ions and the shape of molecular ions are of prime importance. More details are given in chapter 6.

4.4 Lattice Energy

Definition: Lattice energy is the energy required to disassemble one mole of a crystalline compound at a temperature of 0 K in such a way that its components are moved to positions which are infinitely far apart.

In this sense, components are taken to be:

- for molecular compounds: the molecules
- for ionic compounds: the ions
- for metals: the atoms
- for pure elements, excluding molecular species such as H_2, N_2, S_8 etc.: the atoms

For compounds that cannot be assigned uniquely to one of these substance classes, the question of how lattice energy should be defined remains open. Should SiO_2 be decomposed into Si and O atoms or into Si^{4+} and O^{2-} ions? For polar compounds like SiO_2, lattice energy values given in the literature usually refer to a decomposition into ions. Values calculated under this assumption should be considered with caution as the neglected covalent bonds are of considerable importance. Even in the case of a separation into ions conditions are not always clear: should Na_2SO_4 be separated into Na^+ and SO_4^{2-} ions or into Na^+, S^{6+} and O^{2-} ions?

Lattice Energy of Molecular Compounds

The lattice energy E of a molecular compound is equal to the energy of sublimation at 0 K. This energy cannot be measured directly, but it is equal to the enthalpy of sublimation at a temperature T plus the thermal energy needed to warm the sample from 0 K to this temperature, minus RT. RT is the amount of energy required to expand one mole of a gas at a temperature T to an infinitely small pressure. These amounts of energy can be measured and therefore the lattice energy can be determined experimentally in this case.

The following four forces acting between molecules contribute to the lattice energy of a crystal consisting of molecules:

1. The dispersion forces (LONDON forces) which always are attractive.

2. The repulsion due to the interpenetration of the electron shells of atoms that come together too closely.

3. For molecules with polar bonds, i.e. for molecules having the character of dipoles or multipoles, the electrostatic interaction between the dipoles or multipoles.

4. The zero point energy which always is present even at the absolute zero point of temperature.

The zero point energy follows from quantum theory, according to which atoms do not cease to vibrate at the absolute zero point. For a DEBYE solid (that is a homogeous body of N equal particles) the zero point energy is

$$E_0 = N\frac{9}{8}h\nu_{max}$$

ν_{max} is the frequency of the highest occupied vibrational state in the crystal. For molecules with a very small mass and for molecules that are being held together via hydrogen bridges, the zero point energy makes a considerable contribution. For H_2 and He it even amounts to the predominant part of the lattice energy. For H_2O it contributes about 30 %, and for N_2, O_2 and CO about 10 %. For larger molecules the contribution of the zero point energy is marginal.

The dispersion force between two atoms results in the dispersion energy E_D, which is approximately proportional to r^{-6}, r being the distance between the atoms:

$$E_D = -\frac{C}{r^6}$$

The dispersion energy between two molecules results approximately from the sum of the contributions of atoms of one molecule to atoms of the other molecule.

For the repulsion energy E_A between two atoms that come together too closely an exponential function is usually taken:

$$E_A = B\,e^{-\alpha r}$$

Another, equally appropriate approximation is:

$$E_A = B'r^{-n}$$

with values for n ranging between 5 and 12 (BORN repulsion term).

For molecules with low polarity like hydrocarbons, electrostatic forces only have a minor influence. Molecules with highly polar bonds behave as dipoles or multipoles and exhibit corresponding interactions. For instance, hexahalide molecules like WF_6 or WCl_6 are multipoles, the halogen atoms bearing a negative partial charge $-q$, while the metal atom has a positive charge $+6q$; the partial charge q has some value between zero and one, but its exact amount is not usually known. Although the forces exerted by a multipole only have appreciable influence on close-lying molecules, they can contribute significantly to the lattice energy. The electrostatic or Coulomb energy E_C for two interacting atoms having the charges q_i and q_j is:

$$E_C = \frac{1}{4\pi\varepsilon_0}\frac{q_iq_je^2}{r} \tag{1}$$

q_i, q_j being given in units of the electrical unit charge $e = 1.6022 \times 10^{-19}$ C. $\varepsilon_0 =$ electric field constant $= 8.859\times10^{-12}$ $C^2J^{-1}m^{-1}$.

Altogether, the lattice energy E can thus be calculated according to the following approximation:

$$E = -N_A \sum (E_D + E_A + E_C + E_0)$$

$$= -N_A \sum_{i,j} \left[-C_{ij} r_{ij}^{-6} + B_{ij} \exp(-\alpha_{ij} r_{ij}) + \frac{q_i q_j e^2}{4\pi\varepsilon_0 r_{ij}} + \frac{9}{8} h\nu_{max} \right] \qquad (2)$$

The choice of the signs gives a positive value for the lattice energy, corresponding to its definition as energy that has to be supplied in order to destroy the crystal lattice. The atoms of *one* molecule are counted with the index i, while *all* atoms of all other molecules in the crystal are counted with the index j. This way the interaction energy of one molecule with all other molecules is calculated. The lattice energy per mole results from multiplication by AVOGADRO's number N_A. r_{ij} is the distance between the atoms i and j, q_i and q_j are their partial charges in units of the electric unit charge. B_{ij}, α_{ij} and C_{ij} are parameters that have to be determined experimentally. As the contributions of the terms in equation (2) decrease with growing distances r_{ij}, a sufficient accuracy can be obtained by considering only atoms up to some upper limit for r_{ij}. The summation can then be performed rather quickly with a computer.

Values for the partial charges of atoms can be derived from quantum mechanical calculations, from the molecular dipole moments and from rotation–vibration spectra. However, in most cases they are not well known and as a consequence, the contribution of the Coulomb energy cannot be calculated precisely, so that no reliable lattice energy calculations are possible. Parameters B_{ij}, α_{ij} and C_{ij} are known for the interactions between atoms of some elements such as H, C, N, O, F and Cl. Examples of the resulting potential functions are shown in Fig. 12. The minimum point in each graph corresponds to the interatomic equilibrium distance between *two single* atoms. In a crystal shorter distances result because a molecule contains several atoms and thus several attractive atom–atom forces are active between two molecules, and because attractive forces with further surrounding molecules cause an additional compression.

Fig. 12
Potential functions for the interaction energies due to repulsive and to dispersion forces between two atoms as a function of the interatomic distance

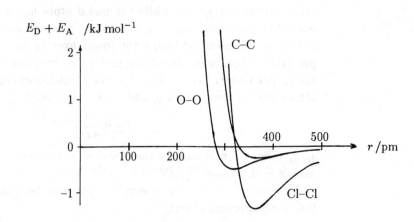

Generally, increasing molecular size, heavier atoms and more polar bonds contribute to an increased lattice energy of a molecular crystal. Typical values are: argon 7.7 kJ mol^{-1}; krypton 11.1 kJ mol^{-1}; benzene 49.7 kJ mol^{-1}.

The Lattice Energy of Ionic Compounds

In a molecule the partial charges of all atoms add up to zero and therefore the repulsive and the attractive electrostatic forces between two molecules are more or less balanced. Only the uneven distribution of the charges causes a certain electrostatic contribution to the lattice energy. For a polyatomic ion, however, the partial charges do not add up to zero, but to the value of the ionic charge. As a consequence, strong electrostatic interactions are present between ions, their contributions amounting to the main part of the lattice energy. Numerical values for sodium chloride illustrate this:

$$-N_A \sum q_i q_j e^2 / (4\pi\varepsilon_0 \, r_{ij}) = 867 \qquad \text{Coulomb energy}$$
$$-N_A \sum B_{ij} \exp(-\alpha_{ij} r_{ij}) = -92 \qquad \text{repulsion energy}$$
$$N_A \sum C_{ij} r_{ij}^{-6} = 18 \qquad \text{dispersion energy}$$
$$-2N_A \tfrac{9}{8} h\nu_{\max} = -6 \qquad \text{zero point energy}$$
$$E = \overline{787} \quad \text{kJ mol}^{-1}$$

The calculation of the lattice energy can be performed with the aid of equation (2). In the case of monoatomic ions the charges q take the values of the ionic charges. Crystals consisting of monoatomic ions like Na$^+$ or Cl$^-$ have simple and symmetrical structures, which are useful for the summation according to equation (2). Let us take the structure of NaCl as an example. If we designate the shortest distance Na$^+$—Cl$^-$ in the crystal by R, then all other interionic distances can be given as multiples of R. Which multiples occur follows from simple geometrical considerations on the basis of the structure model of NaCl (cf. Fig. 14, p. 38). For a single Na$^+$ ion within the crystal the Coulomb energy turns out to be:

$$
\begin{aligned}
E_C &= \frac{N_A}{4\pi\varepsilon_0} \sum_j \frac{q_1 q_j e^2}{r_{1j}} \\
&= \frac{N_A e^2}{4\pi\varepsilon_0 R} \left(-\frac{6}{1} + \frac{12}{\sqrt{2}} - \frac{8}{\sqrt{3}} + \frac{6}{2} - \frac{24}{\sqrt{5}} + \cdots \right)
\end{aligned}
\tag{3}
$$

The terms in brackets result as follows:

1. -6, because the Na$^+$ ion is surrounded by 6 Cl$^-$ ions with charge -1 at a distance $r = R$ in a first sphere.

2. $+12/\sqrt{2}$, because the Na$^+$ ion is surrounded by 12 Na$^+$ ions with charge $+1$ at a distance $R\sqrt{2}$ in a second sphere.

3. $-8/\sqrt{3}$, because there are 8 Cl$^-$ ions at a distance $R\sqrt{3}$.

Extending the series given in brackets to infinite length, it converges to the value of $-A = -1.74756$. For short, we can write:

$$E_C = -\frac{N_A e^2}{4\pi\varepsilon_0 R} A \tag{4}$$

The quantity A is called MADELUNG constant. By comparison of equation (1) with equation (4) we can see that more energy is liberated by combining the ions into a crystal than by forming one mole of separate ion pairs (assuming equal interionic distances $r = R$; as a matter of fact, for a single ion pair $r < R$, so that the energy gain actually does not quite attain the factor A).

The same constant A can also be used when ions with higher charges are involved, as long as the structre type is the same:

$$E_C = -|q_1||q_2|\frac{N_A e^2}{4\pi\varepsilon_0 R} A$$

$|q_1|$ and $|q_2|$ are the absolute values of the ionic charges as multiples of the electric unit charge. The MADELUNG constant is independent of the ionic charges and of the lattice dimensions, but it is valid only for one specific structure type. Table 1 lists the values for some simple structure types.

Table 1 MADELUNG constants for some structure types

structure type	A	structure type	A
CsCl	1.76267	CaF_2	5.03879
NaCl	1.74756	TiO_2 (rutile)	4.816
ZnS (wurtzite)	1.64132	$CaCl_2$	4.730
ZnS (sphalerite)	1.63805	$CdCl_2$	4.489
		CdI_2	4.383

MADELUNG constants only cover the coulombic part of the lattice energy provided that the values of the charges q_1 and q_2 are known. A complete separation of charges between anions and cations yielding integer values for the ionic charges is met quite well only for the alkali metal halides. When some covalent bonding is present, partial charges must be assumed. The magnitudes of these partial charges are not usually known. In this case absolute values for the coulombic part of the lattice energy cannot be calculated. For ZnS, TiO_2, $CdCl_2$ and CdI_2 differing polarities have to be assumed, so that the values listed in Table 1 do not follow the real trend of the lattice energies. Nevertheless, MADELUNG constants are useful quantities; they can serve to estimate which structure type should be favored energetically by a compound when the Coulomb energy is the determining factor.

4.5 Problems

4.1 The densities of some SiO_2 modifications are: α-quartz 2.65 $g\,cm^{-3}$, β-quartz 2.53 $g\,cm^{-3}$, β-tridymite 2.27 $g\,cm^{-3}$, β-cristobalite 2.33 $g\,cm^{-3}$, vitreous SiO_2 2.20 $g\,cm^{-3}$. Should it be possible to convert β-cristobalite to some of the other modifications by applying pressure?

4.2 Silica glass is formed when molten SiO_2 is cooled rapidly. It experiences slow crystallization. Will the rate of crystallization be higher at room temperature or at 1000 °C?

4.3 BeF_2, like quartz, has a polymer structure with F atoms linking tetrahedrally coordinated Be atoms; BF_3 is monomer. When cooling the liquid down to solidification, which of the two is more likely to form a glass?

4.4 Derive the first four terms of the series to calculate the MADELUNG constant for CsCl (Fig. 14).

4.5 Calculate the contribution of the Coulomb energy to the lattice energy of:

(a) CsCl, $R = 356$ pm;

(b) CaF_2, $R = 236$ pm;

(c) BaO (NaCl type), $R = 276$ pm.

5 The Effective Size of Atoms

According to wave machanics the electron density in an atom decreases asymptotically toward zero with increasing distance from the atomic center. An atom therefore has no definite size. When two atoms approach each other, interaction forces between them become more and more effective.

Attractive are:

- The ever present dispersion force (LONDON attraction).
- Electronic interactions with the formation of bonding molecular orbitals.
- Electrostatic forces between the charges of ions or the partial charges of atoms having opposite signs.

Repulsive are:

- The electrostatic forces between ions or partially charged atoms having charges of the same sign.
- The interpenetration of closed electron shells of atoms (resulting in antibonding states) and the electrostatic repulsion between their atomic nuclei when the atoms get too close to one another.

The effectiveness of these forces differs and, furthermore, they change to a different degree as a function of the interatomic distance. The last-mentioned repulsion force is by far the most effective at short distances, but its range is rather restricted; at somewhat bigger distances the other forces dominate. At some definite interatomic distance attractive and repulsive forces are balanced. This equilibrium distance corresponds to the minimum in a graph in which the potential energy is plotted as a function of the atomic distance ("potential curve", cf. Fig. 12, p. 26).

The equilibrium distance that always occurs between atoms conveys the impression of atoms being spheres of a definite size. As a matter of fact, in many cases atoms can be treated as if they were more or less hard spheres.

Since the attractive forces between the atoms differ depending on the type of bonding forces, for every kind of atom several different sphere radii have to be assigned according to the bonding types. From experience we know that for one specific kind of bonding the atomic radius of an element has a fairly constant value. We distinguish the following radius types: VAN DER WAALS radii, metallic radii, several ionic radii depending on the ionic charges, and covalent radii for single, double and triple bonds. Furthermore, the values vary depending on coordination numbers; the larger the coordination number, the bigger is the radius.

5.1 Van der Waals Radii

In a crystalline compound consisting of molecules, the molecules usually are packed as close as possible, but with atoms of neighboring molecules not coming closer than the sums of their VAN DER WAALS radii. The shortest commonly observed distance between atoms of the same element in adjacent molecules is taken to calculate the VAN DER WAALS radius for this element. Some values are listed in Table 2. A more detailed study reveals that covalently bonded atoms are not exactly spherical. For instance, a halogen atom bonded to a carbon atom is flattened to some degree, i.e. its VAN DER WAALS radius is shorter in the direction of the extension of the C–halogen bond than in directions perpendicular to this bond (cf. Table 2). If the covalent bond is more polar, as in metal halides, then the deviation from the spherical form is less pronounced.

Distances that are shorter than the sums of the corresponding listed values of the VAN DER WAALS radii occur when there exist special attractive forces. For instance, in a solvated ion the distances between the ion and atoms of the solvent molecules cannot be calculated with the aid of VAN DER WAALS radii. The same applies in the presence of hydrogen bonding.

Table 2 Van der Waals radii /pm

H 120	spherical approximation [79]		He 140	
C 170	N 155	O 152	F 147	Ne 154
Si 210	P 180	S 180	Cl 175	Ar 188
Ge	As 185	Se 190	Br 185	Kr 202
Sn	Sb 200	Te 206	I 198	Xe 216

flattened atoms bonded to C [80]

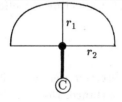

	r_1	r_2		r_1	r_2		r_1	r_2
N	160	160	O	154	154	F	130	138
			S	160	203	Cl	158	178
			Se	170	215	Br	154	184
H	101	126				I	176	213

5.2 Atomic Radii in Metals

The degree of cohesion of the atoms in metals is governed by the extent to which occupation of bonding electron states outweighs antibonding states in the electronic energy bands (cf. section 9.7). Metals belonging to groups in the left part of the periodic table have few valence electrons; the numbers of

Table 3 Atomic radii in metals /pm. All values refer to coordination number 12, except for the alkali metals (c.n. 8), Ga (c.n. 1+6), Sn (c.n. 4+2), Pa (c.n. 10), U, Np and Pu

Li 152	Be 112												
Na 186	Mg 160											Al 143	
K 230	Ca 197	Sc 162	Ti 146	V 134	Cr 128	Mn 137	Fe 126	Co 125	Ni 125	Cu 128	Zn 134	Ga 135	
Rb 247	Sr 215	Y 180	Zr 160	Nb 146	Mo 139	Tc 135	Ru 134	Rh 134	Pd 137	Ag 144	Cd 151	In 167	Sn 154
Cs 267	Ba 222	La 187	Hf 158	Ta 146	W 139	Re 137	Os 135	Ir 136	Pt 139	Au 144	Hg 151	Tl 171	Pb 175

Ce 182	Pr 182	Nd 182	Pm 181	Sm 180	Eu 204	Gd 179	Tb 178	Dy 177	Ho 176	Er 175	Tm 174	Yb 193	Lu 174
Th 180	Pa 161	U 156	Np 155	Pu 159	Am 173	Cm 174	Bk 170	Cf 186	Es 186	Fm	Md	No	Lr

occupied bonding energy states are low. Metals in the right part of the periodic table have many valence electrons; a fraction of them has to be accomodated in antibonding states. In both cases we have relatively weak metallic bonding. When many bonding but few antibonding states are occupied, the resulting bond forces between the metal atoms are large. This is valid for the metals belonging to the central part of the block of transition elements. Atomic radii in metals therefore decrease from the alkali metals up to the metals of the sixth to eighth transition groups, and then they increase. Superimposed on this sequence is the general tendency of decreasing atomic sizes observed in all periods from the alkali metals to the noble gases, which is due to the increasing nuclear charge (Table 3). For intermetallic compounds the ratio of the total number of available valence electrons to the number of atoms (the "valence electron concentration") is a decisive factor affecting the effective atomic size.

5.3   Covalent Radii

Covalent radii are derived from the observed distances between covalently bonded atoms of the same element. For instance, the C–C bond length in diamond and in alkanes is 154 pm; half of this value, 77 pm, is the covalent radius for a single bond at a carbon atom having coordination number 4 (sp^3 C atom). In the same way we calculate the covalent radii for Cl (100 pm) from the Cl–Cl distance in a Cl_2 molecule, for O (73 pm) from the O–O distance in H_2O_2 and for Si (118 pm) from the bond distance in elemental silicon. If we add the covalent radii for C and Cl, we obtain $77 + 100 = 177$ pm; this value corresponds rather well to the distances observed in C–Cl compounds. However, if we add the covalent radii for Si and O, $118 + 73 = 191$ pm, the obtained value does not agree satisfactorily with the distances observed in SiO_2 (158 to 162 pm). Generally we must state: the more polar a bond is, the more its length deviates to lower values compared to the sum of the covalent radii. To take this into account, SHOMAKER and STEVENSON derived the following correction formula:

$$d(AX) = r(A) + r(X) - c|x(A) - x(X)|$$

$d(AX)$ = bond distance, $r(A)$ und $r(X)$ = covalent radii of the atoms A and X, $x(A)$ and $x(X)$ = electronegativities of A and X.

The correction parameter c depends on the atoms concerned and has values between 2 and 9 pm. For C–X bonds no correction is necessary when X is an element of the 5th, 6th or 7th main groups, except for N, O and F. The influence of bond polarity also shows up in the fact that the bond lengths depend on the oxidation states; for instance, the P–O bonds in P_4O_6 (164 pm) are longer than in P_4O_{10} (160 pm; sum of the covalent radii 183 pm). Deviations of this kind are larger for "soft" atoms, i.e. for atoms that can be polarized easily.

Calculated bond lengths also are uncertain when it is not known to exactly what degree multiple bonding is present, what influence lone electron pairs exercise on adjacent bonds, and to what extent the ionic charge and the coordination number have an effect. The range of Cl–O bond lengths illustrates this: HOCl 170 pm, ClO_2^- 156 pm, ClO_3^- 149 pm, ClO_4^- 143 pm, $HOClO_3$ one at 164 and three at 141 pm, ClO_2 147 pm, ClO_2^+ 131 pm. Problems related to bond lengths are also dealt with on pp. 46, 55 and 57.

5.4 Ionic Radii

The shortest cation–anion distance in an ionic compound corresponds to the sum of the ionic radii. This distance can be determined experimentally. However, there is no straightforward way to obtain values for the radii themselves. Data taken from thoroughfully performed X-ray diffraction experiments allow the calculation of the electron density in the crystal; the point having the minimum electron density along the connection line between a cation and an adjacent anion can be taken as the contact point of the ions. As shown in the example of lithium fluoride in Fig. 13, the ions in the crystal show certain deviations from the spherical shape, i.e. the electron shell is polarized. This indicates the presence of some degree of covalent bonding, which can be interpreted as a partial backflow of electron density from the anion to the cation. The electron density minimum therefore does not necessarily represent the ideal place for the limit between cation and anion.

The commonly used values for ionic radii are based on an arbitrarily assigned standard radius for a certain ion. This way, a consistent set of radii for other ions can be derived. Several tables have been published, each of which was derived using a different standard value for one ionic radius (ionic radii by GOLD-SCHMIDT, PAULING, AHRENS, SHANNON). The values by SHANNON are based

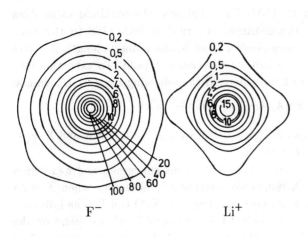
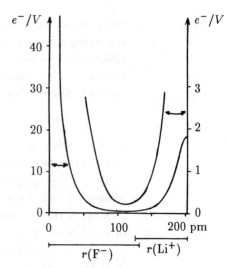

Fig. 13

Experimentally determined electron density in crystalline lithium fluoride (multiples of $10^{-6} e^-/pm^3$; [81]). Left: plane of intersection through adjacent ions; right: electron density along the connecting line F^-—Li^+ and marks for the ionic radii values according to Table 4

on a critical evaluation of experimentally determined interatomic distances and on the standard radius of 140 pm for the O^{2-} ion with sixfold coordination. They are listed in Tables 4 and 5.

Ionic radii can also be used when considerable covalent bonding is involved. The higher the charge of a cation, the more it has a polarizing effect on a neighboring anion, i.e. the covalent character of the bond increases. Nevertheless, arithmetically one can still assume a constant radius for the anion and assign a radius for the cation that will yield the correct interatomic distance. A value like $r(Nb^{5+}) = 64$ pm therefore does not imply the existence of a Nb^{5+} ion with this radius, but means that in niobium(V) compounds the bond length between an Nb atom and a more electronegative atom X can be calculated as the sum of $r(Nb^{5+})$ plus the anionic radius of X. However, the values are not completely independent of the nature of X; for instance, the values given in Table 5 cannot be used readily for sulfur compounds; for this purpose another set of slightly different ionic radii has been derived [83].

Ionic radii of soft (easily polarized) ions depend on the counter-ion. The H^- ion is an example; its radius is 130 pm in MgH_2, 137 pm in LiH, 146 pm in NaH and 152 pm in KH.

The ionic radii listed in Tables 4 and 5 in most cases apply to ions which have coordination number 6. For other coordination numbers slightly different values have to be taken. For every unit by which the coordination number increases or decreases, the ionic radius increases or decreases by 1.5 to 2 %. For coordination number 4 the values are approximately 4 % smaller, and for coordination number 8 about 3 % greater than for coordination number 6. The reason for this is the mutual repulsion of the coordinated ions, the effect of which increases when more of them are present. The size of the coordinated ions also has some influence: a cation that is surrounded by six small anions appears to be slightly smaller than the same cation surrounded by six large anions because in the latter case the anions repel each other more. To account for this, a correction function was derived by PAULING [23]. When covalent bonding is involved, the ionic radii depend to a larger extent on the coordination number. For instance, increasing the coordination number from 6 to 8 entails an increase of the ionic radii of lanthanoid ions of about 13 %, and for Ti^{4+} and Pb^{4+} of about 21 %. An ionic radius decrease of 20 to 35 % is observed when the coordination number of a transition element decreases from 6 to 4.

Table 4 Ionic radii for main group elements according to SHANNON [82], based on $r(O^{2-}) = 140$ pm. Numbers with signs: oxidation states. All values refer to coordination number 6 (except c.n. 4 for N^{3-}).

H	Li	Be	B	C	N	O	F
-1 ~150	$+1$ 76	$+2$ 45	$+3$ 27	$+4$ 16	-3 146 $+3$ 16	-2 140	-1 133
	Na	Mg	Al	Si	P	S	Cl
	$+1$ 102	$+2$ 72	$+3$ 54	$+4$ 40	$+3$ 44 $+5$ 38	-2 184 $+6$ 29	-1 181
	K	Ca	Ga	Ge	As	Se	Br
	$+1$ 138	$+2$ 100	$+3$ 62	$+2$ 73 $+4$ 53	$+3$ 58 $+5$ 46	-2 198 $+4$ 50	-1 196
	Rb	Sr	In	Sn	Sb	Te	I
	$+1$ 152	$+2$ 118	$+3$ 80	$+2$ 118 $+4$ 69	$+3$ 76 $+5$ 60	-2 221 $+4$ 97 $+6$ 56	-1 220 $+5$ 95 $+7$ 53
	Cs	Ba	Tl	Pb	Bi	Po	
	$+1$ 167	$+2$ 135	$+1$ 150 $+3$ 89	$+2$ 119 $+4$ 78	$+3$ 103 $+5$ 76	$+4$ 94 $+6$ 67	

5.5 Problems

5.1 In the following tetrahedral molecules the bond distances are:
SiF_4 155 pm; $SiCl_4$ 202 pm; SiI_4 243 pm.
Calculate the halogen–halogen distances and compare them with the VAN DER WAALS distances. What do you conclude?

5.2 Use ionic radii to calculate expected bond lengths for:
Molecules WF_6, WCl_6, PCl_6^-, PBr_6^-, SbF_6^-, MnO_4^{2-};
Solids (metal atom has c.n. 6) TiO_2, ReO_3, EuO, $CdCl_2$.

Table 5 Ionic radii for transition elements according to SHANNON [82], based on $r(O^{2-}) = 140$ pm. Numbers with signs: oxidation states; ls = low spin, hs = high spin; roman numerals: coordination numbers if other than 6.

	Sc	Ti	V	Cr	Mn	Fe	Co	Ni	Cu	Zn	
+2				ls 73	ls 67	ls 61	ls 65		+1 77		+2
+2		86	79	hs 80	hs 83	hs 78	hs 75	69	73	74	+2
+3	75	67	64	62	ls 58	ls 55	ls 55	ls 56	ls 54		+3
+3					hs 65	hs 65	hs 61	hs 60			+3
+4		61	58	55	53	59	hs 53	ls 48			+4
+5			54	49	IV 26						+5
+6				44	IV 25	IV 25					+6

	Y	Zr	Nb	Mo	Tc	Ru	Rh	Pd	Ag	Cd	
+1									115		+1
+2								86	94	95	+2
+3	90		72	69		68	67	76	75		+3
+4		72	68	65	65	62	60	62			+4
+5			64	61	60	57	55				+5
+6				59							+6

	La	Hf	Ta	W	Re	Os	Ir	Pt	Au	Hg	
+1									137	119	+1
+2								80		102	+2
+3	103		72				68		85		+3
+4		71	68	66	63	63	63	63			+4
+5			64	62	58	58	57	57	57		+5
+6				60	55	55					+6

	Ac
+3	112

	Ce	Pr	Nd	Pm	Sm	Eu	Gd	Tb	Dy	Ho	Er	Tm	Yb	Lu
+2						117			107			103	102	
+3	101	99	98	97	96	95	94	92	91	90	89	88	87	86
+4	87	85						76						

	Th	Pa	U	Np	Pu	Am	Cm	Bk	Cf	Es	Fm	Md	No	Lr
+3		104	103	101	100	98	97	96	95					
+4	94	90	89	87	86	85	85	83	82					
+5		78	76	75	74									
+6			73	72	71									

6 Ionic Compounds

6.1 Radius Ratios

In an energetically favorable packing of cations and anions only anions are directly adjacent to a cation and vice versa. This way, the attractive forces between ions of opposite charges outweigh the repulsive forces between ions of like charges. Packing many ions together into a crystal frees an amount of energy which is larger by the factor A than in the formation of separate ion pairs (assuming equal interionic distances R). A is the MADELUNG constant discussed in section 4.4 (p. 27), which has a definite value for a given crystal structure type. One might now think that the structure type having the largest MADELUNG constant for a given stoichiometry should always be favored. However, this is not the case.

The stability of a certain structure type depends essentially on the relative sizes of cations and anions. Even with a larger MADELUNG constant a structure type can be less stable than another structure type in which cations and anions can approach each other more closely; this is so because the lattice energy also depends on the interionic distances (cf. equation (4), p. 28). The relative size of the ions is quantified by the *radius ratio* r_M/r_X, r_M being the cation radius and r_X the anion radius. In the following the ions are taken to be hard spheres having specific radii.

For compounds of the composition MX (M = cation, X = anion) the CsCl type has the largest MADELUNG constant. In this structure type a Cs^+ ion is in contact with eight Cl^- ions in a cubic arrangement (Fig. 14). The Cl^- ions have no contact with one another. With cations smaller than Cs^+ the Cl^- ions come closer together and when the radius ratio has the value of $r_M/r_X = 0.732$, the Cl^- ions are in contact with each other. When $r_M/r_X < 0.732$, the Cl^- ions remain in contact, but there is no more contact between anions and cations. Now another structure type is favored: its MADELUNG constant is indeed smaller, but it again allows contact of cations with anions. This is achieved by the smaller coordination number 6 of the ions that is fulfilled in the NaCl type (Fig. 14). When the radius ratio becomes even smaller, the sphalerite (zinc blende) or the wurtzite type should occur, in which the ions only have the coordination number 4 (Fig. 14; sphalerite and wurtzite are two modifications of ZnS).

Fig. 14
The three most important structure types for ionic compounds of composition MX. Compared to their effective sizes, the ions have been drawn to a smaller scale

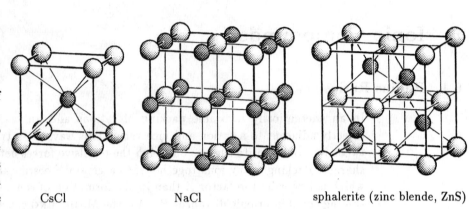

CsCl   NaCl   sphalerite (zinc blende, ZnS)

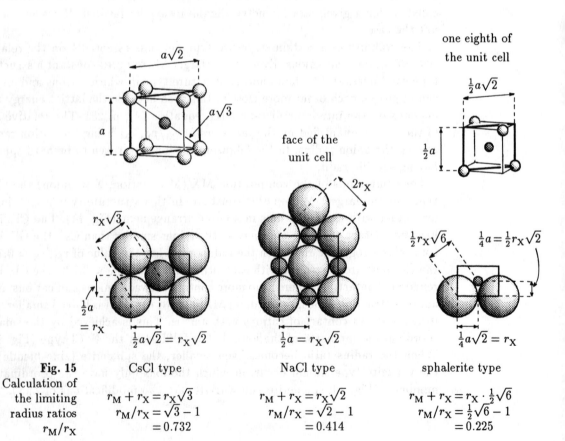

Fig. 15
Calculation of the limiting radius ratios r_M/r_X

CsCl type

$r_M + r_X = r_X\sqrt{3}$
$r_M/r_X = \sqrt{3} - 1$
$= 0.732$

NaCl type

$r_M + r_X = r_X\sqrt{2}$
$r_M/r_X = \sqrt{2} - 1$
$= 0.414$

sphalerite type

$r_M + r_X = r_X \cdot \frac{1}{2}\sqrt{6}$
$r_M/r_X = \frac{1}{2}\sqrt{6} - 1$
$= 0.225$

The geometric considerations leading to the following values are outlined in Fig. 15:

r_M/r_X	coordination number and polyhedron		structure type
> 0.732	8	cube	CsCl
0.414 to 0.732	6	octahedron	NaCl
< 0.414	4	tetrahedron	sphalerite

The purely geometrical approach we have considered so far still is too simple. The really determining factor is the lattice energy, the calculation of which is somewhat more complicated. If we only take into account the electrostatic part of the lattice energy, then the relevant magnitude in equation (4) is the ratio A/R (A = MADELUNG constant, R = shortest cation–anion distance). Fig. 16 shows how the electrostatic part of the lattice energy depends on the radius ratio for chlorides. The transition of the NaCl type to the sphalerite type is to be expected at the crossing point of the curves at $r_M/r_X \approx 0.3$ instead of $r_M/r_X = 0.414$. The transition of the NaCl type to the CsCl type is to be expected at $r_M/r_X \approx 0.71$. The curves were calculated assuming hard Cl^- ions with $r_{Cl^-} = 181$ pm. If, in addition, we take into account the increase of the ionic radius for an increased coordination number, then we obtain the dotted line in Fig. 16 for the CsCl type. As a consequence, the CsCl type should not occur at all, as the dotted line always runs below the line for the NaCl type. However, the CsCl type does occur when heavy ions are involved; this is due to the dispersion energy, which has a larger contribution in the case of the CsCl type. Table 6 lists the structure types actually observed for the alkali halides.

Twelve anions can be arranged around a cation when the radius ratio is 1.00. However, unlike to the three structure types considered so far, geometrically the coordination number 12 does not allow for any arrangement which has cations

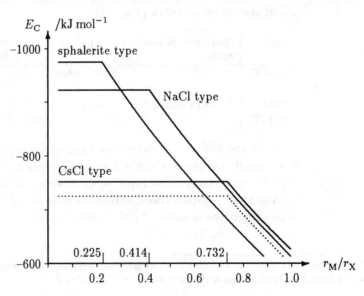

Fig. 16
The electrostatic part of the lattice energy for chlorides crystallizing in the CsCl, NaCl and sphalerite type as a function of the radius ratio

Table 6 Radius ratios and observed structure types for the alkali halides

	Li	Na	K	Rb	Cs	
F	0.57	0.77	0.96*	0.88*	0.80*	
Cl	0.42	0.56	0.76	0.84	0.92	
Br	0.39	0.52	0.70	0.78	0.85	CsCl
I	0.35	0.46	0.63	0,69	0.76	type
		NaCl type				

* r_X/r_M

surrounded only by anions and anions only by cations simultaneously. This kind of coordination therefore does not occur among ionic compounds. When r_M/r_X becomes larger than 1, as for RbF and CsF, the relations are reversed: in this case the cations are larger than the anions and the contacts among the cations determine the limiting radius ratios; the same numerical values and structure types apply, but the inverse radius ratios have to be taken, i.e. r_X/r_M.

The sphalerite type is unknown for truly ionic compounds because there exists no pair of ions having the appropriate radius ratio. However, it is well-known for compounds with considerable covalent bonding even when the sphalerite type is not to be expected according to the the relative sizes of the atoms in the sense of the above-mentioned considerations. Examples are CuCl, AgI, ZnS, SiC, and GaAs. We focus in more detail on this structure type in chapter 11.

In the structure types for compounds MX so far considered, both anions and cations have the same coordination numbers. In compounds MX_2 the coordination number of the cations must be twice that of the anions.* The geometrical considerations concerning the relations of radius ratios and coordination polyhedra are the same. First of all, two structure types fulfil the conditions and are of special importance (Fig. 17):

| r_M/r_X | coordination number and polyhedron | | structure | examples |
	cation	anion	type	
> 0.732	8 cube	4 tetrahedron	fluorite (CaF_2)	SrF_2, BaF_2, ThO_2, $SrCl_2$, $BaCl_2$
0.414 to 0.732	6 octahedron	3 triangle	rutile (TiO_2)	MgF_2, FeF_2, ZnF_2, SnO_2

When the positions of cations and anions are interchanged, the same structure types result for the CsCl, NaCl and sphalerite type. In the case of the fluorite type the interchange also involves an interchange of the coordination numbers, i.e. the anions obtain coordination number 8 and the cations 4. This structure type sometimes is called "anti-fluorite" type; it is known for the alkali oxides (Li_2O, ... , Rb_2O).

---

*When anions with different coordination numbers are present, their mean value has to be doubled. For example: in SrI_2 there exist anions with c.n. 3 and 4; Sr^{2+} has c.n. 7

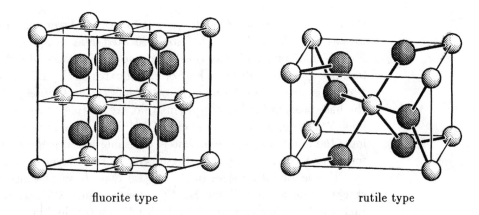

Fig. 17
Fluorite and
rutile type

fluorite type   rutile type

The structure types discussed so far have a favorable arrangement of cations
and anions and are well-suited for ionic compounds consisting of spherical ions.
However, their occurrence is by no means restricted to ionic compounds. The
majority of their representatives are found among compounds with considerable
covalent bonding and among intermetallic compounds.

6.2   Ternary Ionic Compounds

When three different kinds of spherical ions are present, their relative sizes are
also an important factor that controls the stability of a structure. The PbFCl
type is an example having anions packed with different densities according to
their sizes. As shown in Fig. 18, the Cl^- ions form a layer with a square pattern.
On top of that there is a layer of F^- ions, also with a square pattern, but rotated
through 45°. The F^- ions are situated above the edges of the squares of the Cl^-
layer (dotted line in Fig. 18). With this arrangement the F^-–F^- distances are
smaller by a factor of 0.707 ($= \frac{1}{2}\sqrt{2}$) than the Cl^-–Cl^- distances; this matches

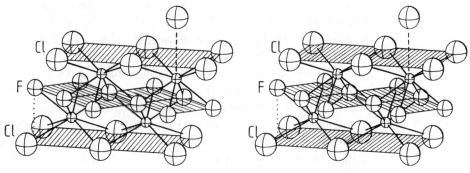

Fig. 18
The PbFCl type
(stereo view).

How to view stereo pictures: the left picture is to be viewed with the left eye, and the right one with the
right eye. It requires some practice to orient the eyes parallel for this purpose. As an aid, one can insert
a cardboard partition between the pictures, so that each eye can see only one picture. Fix your view on
some outstanding feature like the upper right atom until the two pictures merge into one

the ionic radius ratio of $r_{F^-}/r_{Cl^-} = 0.73$. An F^- layer contains twice as many ions as a Cl^- layer. Every Pb^{2+} ion is located in an antiprism having as vertices four F^- and four Cl^- ions that form two square faces of different sizes. Pb^{2+} ions are located under one half of the squares of the F^-; an equal number of Pb^{2+} ions are situated above the other half of the squares, which in turn form the base faces of further antiprisms that are completed by another layer of Cl^- ions. This way, the total number of Pb^{2+} ions is the same as the number of F^- ions; the number of Cl^- ions also is the same because there are two Cl^- layers for every F^- layer. Together, these layers form a layer package that is limited by Cl^- ions on either side. In the crystal these layer packages are stacked with staggered adjacent Cl^- layers. As a consequence, the coordination sphere of each Pb^{2+} ion is completed by a fifth Cl^- ion (dashed in Fig. 18).

Numerous compounds adopt the PbFCl structure. These include, apart from fluoride chlorides, oxide halides MOX (M = Bi, lanthanoids, actinoids; X = Cl, Br, I), hydride halides like CaHCl and many compounds with metallic properties like ZrSiS or NbSiAs.

Further ternary compounds for which the relative sizes of the ions are an important factor for their stability are the perovskites and the spinels, which are discussed in sections 16.4 and 16.6, respectively.

6.3   Compounds with Complex Ions

The structures of ionic compounds comprising complex ions can in many cases be derived from the structures of simple ionic compounds. A spherical ion is substituted by the complex ion and the crystal lattice is distorted in a manner adequate to account for the shape of this ion.

Rod-like ions like CN^-, C_2^{2-} or N_3^- can substitute the Cl^- ions in the NaCl type when all of them are oriented parallel and the lattice is stretched in the corresponding direction. In CaC_2 the acetylide ions are oriented parallel to one of the edges of the unit cell; as a consequence, the symmetry is no longer cubic, but tetragonal (Fig. 19). The hyperoxides KO_2, RbO_2 and CsO_2 as well as peroxides like BaO_2 crystallize in the CaC_2 type. In CsCN and NaN_3 the cyanide and azide ions, respectively, are oriented along one of the space diagonals of the unit cell, and the symmetry is rhombohedral (Fig. 19).

The structure of calcite ($CaCO_3$) can be derived from the NaCl structure by substituting the Cl^- ions for CO_3^{2-} ions. These are oriented perpendicular to one of the space diagonals of the unit cell and require an expansion of the lattice perpendicular to this diagonal (Fig. 20). The calcite type is also encountered among borates (e.g. $AlBO_3$) and nitrates ($NaNO_3$). Another way of regarding this structure is discussed on p. 163.

By substituting the Ca^{2+} ions in the CaF_2 type for $PtCl_6^{2-}$ ions and the F^- ion for K^+ ions, one obtains the K_2PtCl_6 type (Fig. 20). It occurs among numerous hexahalo salts. In this structure type each of a group of four $PtCl_6^{2-}$ ions has one octahedron face in contact with one K^+ ion, which therefore has

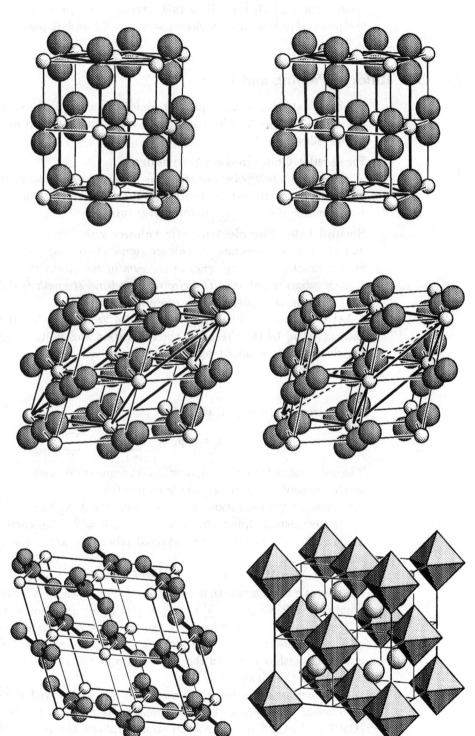

Fig. 19
The structures of CaC_2 and NaN_3. Heavy outlines: tetragonal and rhombohedral (primitive) unit cell, respectively (stereo views)

Fig. 20
The structures of $CaCO_3$ (calcite) and K_2PtCl_6. The section of the calcite structure shown does not correspond to the unit cell (as can be seen from the different orientations of the CO_3^{2-} groups on opposite edges)

coordination number 12. How this structure can be regarded as derived from the perovskite type with a close packing of Cl and K particles is discussed on p. 200.

6.4   The Rules of Pauling and Baur

Important structural principles for ionic crystals, which had already been recognized in part by V. GOLDSCHMIDT, were summmarized by L. PAULING [23] in the following rules.

First rule: Coordination polyhedra
A coordination polyhedron of anions is formed around every cation. The cation-anion distances are determined by the sum of the ionic radii, and the coordination number of the cation by the radius ratio.

Second rule: The electrostatic valence rule
In a stable ionic structure the valence (ionic charge) of each anion with changed sign is exactly or nearly equal to the sum of the electrostatic bond strengths to it from adjacent cations. The electrostatic bond strength is defined as the ratio of the charge on a cation to its coordination number.

Let a be the coordination number of an anion. Of the set of its a adjacent cations, let n_i be the charge on the i-th cation and k_i its coordination number; the electrostatic bond strength of this cation is:

$$s_i = \frac{n_i}{k_i} \tag{5}$$

The charge z_j of the j-th anion is:

$$z_j \approx -p_j = -\sum_{i=1}^{a} s_i = -\sum_{1}^{a} \frac{n_i}{k_i} \tag{6}$$

The rule states that the electrostatic charges in an ionic crystal are balanced locally around every ion as evenly as possible.

For example: Let the cation M^{2+} in a compound MX_2 have coordination number 6. Its electrostatic bond strength is $s = 2/6 = \frac{1}{3}$. The correct charge for the anion, $z = -1$, can only be obtained when the anion has the coordination number $a = 3$.

Let the cation M^{4+} in a compound MX_4 also have coordination number 6; its electrostatic bond strength is $s = 4/6 = \frac{2}{3}$. For an anion X^- having coordination number $a = 2$ we obtain $\sum s_i = \frac{2}{3} + \frac{2}{3} = \frac{4}{3}$; for an anion with $a = 1$ the sum is $\sum s_i = \frac{2}{3}$. For other values of a the resulting p_j deviate even more from the expected value $z = -1$. The most favorable structure will have anions with $a = 2$ and with $a = 1$, and these in a ratio of 1:1, so that the correct value for z results in the mean.

The electrostatic valence rule usually is met rather well by polar compounds, even when considerable covalent bonding is present. For instance, in calcite $(CaCO_3)$ the Ca^{2+} ion has coordination number 6 and thus an electrostatic bond strength of $s(Ca^{2+}) = \frac{1}{3}$. For the C atom, taken as C^{4+} ion, it is $s(C^{4+}) = \frac{4}{3}$.

We obtain the correct value of z for the oxygen atoms, considering them as O^{2-} ions, if every one of them is surrounded by one C and two Ca particles, $z = -[2s(Ca^{2+}) + s(C^{4+})] = -[2 \cdot \frac{1}{3} + \frac{4}{3}] = -2$. This corresponds to the actual structure. $NaNO_3$ and YBO_3 have the same structure; in these cases the rule also is fulfilled when the ions are taken to be Na^+, N^{5+}, Y^{3+}, B^{3+} and O^{2-}. For the numerous silicates no or only marginal deviations result when the calculation is performed with metal ions, Si^{4+} and O^{2-} ions.

The electrostatic valence rule has turned out to be a valuable tool for the distinction of the particles O^{2-}, OH^- and OH_2. Because H atoms often can not be localized by X-ray diffraction, which is the most common method for structure determination, O^{2-}, OH^- and OH_2 can not be distinguished at first. However, their charges must harmonize with the sums p_j of the electrostatic bond strengths of the adjacent cations. For example: kaolinite, $Al_2Si_2O_5(OH)_4$ or "$Al_2O_3 \cdot 2SiO_2 \cdot 2H_2O$", is a layer silicate with Al atoms coordinated octahedrally and Si atoms tetrahedrally; the corresponding electrostatic bond strengths are:

$$s(Al^{3+}) = \tfrac{3}{6} = 0.5 \qquad s(Si^{4+}) = \tfrac{4}{4} = 1.0$$

The atoms in a layer are situated in planes with the sequence O(1)–Al–O(2)–Si–O(3) (cf. Fig. 107, p. 176). The particles O(2), which are shared by octahedra and tetrahedra, have c.n. 3 ($2 \times Al$, $1 \times Si$), the other O particles have c.n. 2. We calculate the following sums of electrostatic bond strengths:

O(1): $p_1 = 2 \cdot s(Al^{3+}) = 2 \cdot 0.5 = 1$
O(2): $p_2 = 2 \cdot s(Al^{3+}) + 1 \cdot s(Si^{4+}) = 2 \cdot 0.5 + 1 = 2$
O(3): $p_3 = 2 \cdot s(Si^{4+}) = 2 \cdot 1 = 2$

Therefore, the OH^- ions must take the O(1) positions and the O^{2-} ions the remaining positions.

Third rule: Linking of polyhedra
An ionic crystal can be described as a set of linked polyhedra. The electrostatic valence rule allows the deduction of the number of polyhedra that share a common vertex, but not how many vertices are common to two adjacent polyhedra. Two shared vertices are equivalent to one shared edge, three or more common vertices are equivalent to a shared face. In the four modifications of TiO_2, rutile, high-presssure TiO_2 (α-PbO_2 type), brookite, and anatase, the Ti atoms have octahedral coordination; as required by the electrostatic valence rule, every coordinated O atom is shared by three octahedra. In rutile and high-pressure TiO_2 every octahedron has two common edges with other octahedra, in brookite there are three and in anatase four shared edges per octahedron. The third rule states in what way the kind of polyhedron linkage affects the stability of the structure: *The presence of shared edges and especially of shared faces in a structure decreases its stability; this effect is large for cations with high charge and low coordination number.*

The stability decrease is due to the electrostatic repulsion between the cations. The centers of two polyhedra are closest to each other in the case of a shared face and they are relatively distant when only one vertex is shared (cf. Fig. 3, p. 6, and Table 20, p. 156).

According to this rule, rutile and, at high-pressures, the modification with the α-PbO$_2$ structure are the most stable forms of TiO$_2$. Numerous compounds crystallize in the rutile type and some in the α-PbO$_2$ type, whereas the brookite and the anatase structure are known only for titanium dioxide.

Exceptions to the rule are observed for compounds with low polarity, i.e. when covalent bonds predominate. Fluorides and oxides (including silicates) usually fulfil the rule, whereas it is rather useless for chlorides, bromides, iodides, and sulfides. For instance, in metal trifluorides like FeF$_3$ octahedra sharing vertices are present, while in most other trihalides octahedra usually share edges or even faces.

In some cases a tendency exactly opposite to the rule is observed, i.e. decreasing stability in the sequence face-sharing > edge-sharing > vertex-sharing. This applies when it is favorable to allow the atoms in the centers of the polyhedra to come close to one another. This is observed when the metal atoms in transition metal compounds dispose of d electrons and tend to form metal–metal bonds. For instance, the tribromides and triiodides of titanium and zirconium form columns consisting of octahedra sharing opposite faces, with metal atoms forming M–M bonds pairwise between adjacent octehedra (cf. Fig. 96, p. 167).

Fourth rule: Linking of polyhedra having different cations

In a crystal containing different cations those with large charge and small coordination number tend not to share polyhedron elements with each other, i.e. they tend to keep as far apart as possible.

Silicates having an O:Si ratio larger than or equal to 4 are orthosilicates, i.e. the SiO$_4$ tetrahedra do not share atoms with each other, but with the polyhedra about the other cations. Examples: olivines, M$_2$SiO$_4$ (M = Mg^{2+}, Fe^{2+}) and garnets, M$_3$M$_2'$[SiO$_4$]$_3$ (M = Mg^{2+}, Ca^{2+}, Fe^{2+}; M' = Al^{3+}, Y^{3+}, Cr^{3+}, Fe^{3+}).

The extended electrostatic valence rules

Two additional rules, put forward by W. H. BAUR [84], deal with the bond distances $d(MX)$ in ionic compounds:

The bond distances $d(MX)$ within the coordination polyhedron of a cation M vary in the same manner as the values p_j corresponding to the anions X,

and

For a given pair of ions the average value of the distances $d(MX)$ within a coordination polyhedron, $\overline{d(MX)}$, is approximately constant and independent of the sum of the p_j values received by all the anions in the polyhedron. The deviation of an individual bond length from the average value is proportional to $\Delta p_j = p_j - \overline{p}$ ($\overline{p}$ = mean value of the p_j for the polyhedron). Therefore, the bond lengths can be predicted from the equation:

$$d(MX(j)) = \overline{d(MX)} + b\Delta p_j \qquad (7)$$

$\overline{d(MX)}$ and b are empirically derived values for given pairs of M and X in a given coordination.

For example: In baddeleyite, a modification of ZrO_2, Zr^{4+} has coordination number 7 in the sense of the formula $ZrO_{3/3}O_{4/4}$; i.e. there are two kinds of O^{2-} ions, $O(1)$ with c.n. 3 and $O(2)$ with c.n. 4. The electrostatic valence strength of a Zr^{4+} ion is:

$$s = \frac{4}{7}$$

For $O(1)$ and $O(2)$ we calculate:

$$O(1): \quad p_1 = 3 \cdot \frac{4}{7} = 1.714 \qquad O(2): \quad p_2 = 4 \cdot \frac{4}{7} = 2.286$$

We expect shorter distances for $O(1)$; the observed mean distances are:

$$d(Zr-O(1)) = 209 \text{ pm} \quad \text{and} \quad d(Zr-O(2)) = 221 \text{ pm}$$

The average values are:

$$\overline{d(ZrO)} = \frac{1}{7}(3 \cdot 209 + 4 \cdot 221) = 216 \text{ pm} \quad \text{and} \quad \overline{p} = \frac{1}{7}(3 \cdot 1.714 + 4 \cdot 2.286) = 2.041$$

With $b = 20.4$ pm the actual distances cán be calculated according to equation (7).

Table 7 lists values for $\overline{d(MX)}$ and b that have been derived from extensive data. They can be used to calculate the bond distances in oxides, usually with deviations of less than ± 2 pm from the actual values.

Table 7 Some average values $\overline{d(MO)}$ and parameters b for the calculation of bond distances in oxides according to equation (7) [84]

bond	ox. state	c.n.	$\overline{d(MO)}$ /pm	b /pm	bond	ox. state	c.n.	$\overline{d(MO)}$ /pm	b /pm
Li–O	+1	4	198	33	Si–O	+4	4	162	9
Na–O	+1	6	244	24	P–O	+5	4	154	13
Na–O	+1	8	251	31	S–O	+6	4	147	13
K–O	+1	8	285	11					
Mg–O	+2	6	209	12	Ti–O	+4	6	197	20
Ca–O	+2	8	250	33	V–O	+5	4	172	16
B–O	+3	3	137	11	Cr–O	+3	6	200	16
B–O	+3	4	148	13	Fe–O	+2	6	214	30
Al–O	+3	4	175	9	Fe–O	+3	6	201	22
Al–O	+3	6	191	24	Zn–O	+2	4	196	18

6.5 Problems

6.1 Use ionic radius ratios (Tables 4 and 5) to decide whether the CaF_2 or the rutile type is more likely to be adopted by: NiF_2, CdF_2, GeO_2, K_2S.

6.2 In garnet, $Mg_3Al_2Si_3O_{12}$, an O^{2-} ion is surrounded by 2 Mg^{2+}, 1 Al^{3+} and 1 Si^{4+} particle. There are cation sites having coordination numbers of 4, 6 and 8. Use PAULING's second rule to decide which cations go in which sites.

6.3 YIG (yttrium iron garnet), $Y_3Fe_5O_{12}$, has the same structure as garnet. Which are the appropriate sites for the Y^{3+} and Fe^{3+} ions? If the electrostatic valence rule is insufficient for you to come to a decision, take ionic radii as an additional criterion.

6.4 In crednerite, $Cu^{[2l]}Mn^{[6o]}O_2^t$, every oxygen atom is surrounded by 1 Cu and 3 Mn. Can the electrostatic valence rule help to decide whether the oxidation states are Cu^+ and Mn^{3+}, or Cu^{2+} and Mn^{2+}?

6.5 Silver cyanate, AgNCO, consists of infinite chains of alternating Ag^+ and NCO^- ions. Ag^+ has c.n. 2 and only one of the terminal atoms of the cyanate group is part of the chain skeleton, being coordinated to 2 Ag^+. Decide with the aid of PAULING's second rule which of the cyanate atoms (N or O) is the coordinated one. (Decompose the NCO^- to N^{3-}, C^{4+} and O^{2-}).

6.6 In $Rb_2V_3O_8$ the Rb^+ ions have coordination number 10; there are two kinds of vanadium ions, V^{4+} with c.n. 5 and V^{5+} with c.n. 4, and four kinds of O^{2-} ions. The mutual coordination of these particles is given in the table, the first value referring to the number of O's per cation, the second one to the number of cations per O (the sums of the first numbers per row and of the second numbers per column correspond to the c.n.):

	O(1)	O(2)	O(3)	O(4)	c.n.
Rb^+	2;4	4;2	1;2	3;3	10
V^{4+}	1;1	4;1	–	–	5
V^{5+}	–	2;1	1;2	1;1	4
c.n.	5	4	4	4	

Calculate the electrostatic bond strengths of the cations and determine how well the electrostatic valence rule is fulfilled. Calculate the expected individual V–O bond distances using data from Table 7 and the values $d(V^{4+}O) = 189$ pm and $b(V^{4+}O) = 36$ pm.

7 Molecular Structures I: Compounds of Main Group Elements

Molecules and molecular ions consist of atoms that are connected by covalent bonds. Aside from few exceptions, molecules and molecular ions only exist when hydrogen or elements of the fourth to seventh main group of the periodic table are involved (the exceptions are molecules such as Li_2 in the gas phase). These elements tend to attain the electron configuration of the noble gas that follows them in the periodic table. For every covalent bond in which one of their atoms participates, it gains one electron. The **8−N rule** holds: *an electron configuration corresponding to a noble gas is attained when the atom takes part in $8 - N$ covalent bonds; N* = main group number = 4 to 7 (except for hydrogen).

Usually a molecule consists of atoms with different electronegativities, and the more electronegative atoms have smaller coordination numbers (we only count covalently bonded atoms as belonging to the coordination sphere of an atom). The more electronegative atoms normally fulfil the $8 - N$ rule; in many cases they are "terminal atoms", i. e. they have coordination number 1. Elements of the second period of the periodic table almost never surpass the coordination number 4 *in molecules*. However, for elements of higher periods this is quite common, the $8 - N$ rule being violated in this case.

The structure of a molecule depends essentially on the covalent bond forces acting between its atoms. In the first place, they determine the *constitution* of the molecule: that is the sequence of the linkage of the atoms with each other. The constitution can be expressed in a simple way by means of the valence bond formula. For a given constitution the atoms arrange themselves in space according to certain principles. These include: atoms not bonded directly with one another may not come too close (repulsion of interpenetrating electron shells); and the valence electron pairs of an atom keep as far apart as possible from each other.

7.1 Valence Shell Electron-pair Repulsion

The structures of numerous molecules can be understood and predicted with the *valence shell electron-pair repulsion theory* (VSEPR theory) by GILLESPIE and NYHOLM [21,22,88]. In the first place, it is applicable to compounds of main group elements. The special aspects concerning transition group elements are dealt with in chapter 8. However, transition group elements having the electron configurations d^0, high-spin d^5, and d^{10} can be treated in the same way as main group elements; the d electrons need not be taken into account in these cases. In order to apply the theory, one first draws a valence bond formula with the correct constitution, including all lone electron pairs. This formula shows how many valence electron pairs are to be considered at an atom. Every electron pair is taken as one unit (orbital). The electron pairs are being attracted by the corresponding atomic nucleus, but they excercise a mutual repulsion. A function proportional to $1/r^n$ can be used to approximate the repulsion energy between two electron pairs; r is the distance between the centers of gravity of their charges, and n has some value between 5 and 12 ($n = 1$ would correspond to a purely electrostatic repulsion, $n = \infty$ would correspond to hard, inpenetrable orbitals).

The next step is to consider how the electron pairs have to be arranged to achieve a minimum energy for their mutual repulsion. If the centers of the

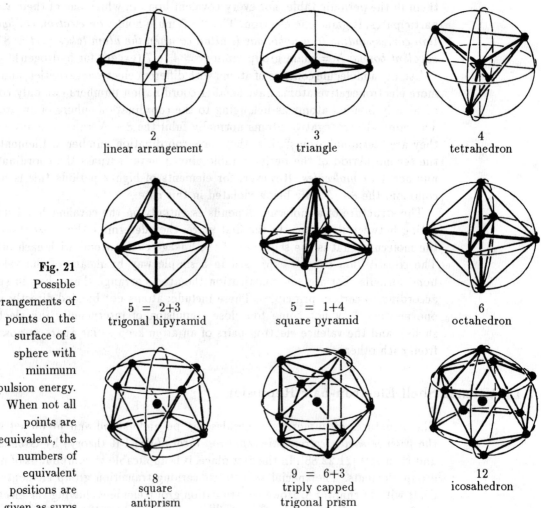

Fig. 21
Possible arrangements of points on the surface of a sphere with minimum repulsion energy. When not all points are equivalent, the numbers of equivalent positions are given as sums

2
linear arrangement

3
triangle

4
tetrahedron

5 = 2+3
trigonal bipyramid

5 = 1+4
square pyramid

6
octahedron

8
square antiprism

9 = 6+3
triply capped trigonal prism

12
icosahedron

charge of all orbitals are equidistant from the atomic nucleus, every orbital can be represented by a point on the surface of a sphere. The problem thus amounts to deduce what distribution of the points on the sphere corresponds to a minimum for the sum $\sum(1/r_i^n)$, covering all distances r_i between the points [89]. As a result, we obtain a definite polyhedron for every number of points (Fig. 21). Only for 2, 3, 4, 6, 8, 9, and 12 points is the resulting polyhedron independent of the value of the exponent n. For five points the trigonal bipyramid is only slightly more favorable than the square pyramid. A model to visualize the mutual arrangement of the orbitals about a common center consists of tightly joined balloons; the pressure in the balloons simulates the value of n.

Molecules having no lone electron pairs at the central atom and having only equal atoms bonded to this atom usually have structures that correspond to the polyhedra shown in Fig. 21.

Some polyhedra have vertices that are not equivalent in the first place. In any case, non-equivalent points always result when the corresponding orbitals belong to bonds with atoms of different elements or when some of the points represent lone electron pairs. In these cases the charge centers of the orbitals have different distances from the atomic center. A charge center that is closer to the atomic nucleus also has shorter distances to the remaining orbitals and thus exerts stronger repulsions on them. In the balloon model an electron pair close to the atomic nucleus corresponds to a larger balloon. This has important consequences for the molecular structure. The following aspects have to be taken into account:

1. A lone electron pair is under the direct influence of only one atomic nucleus, and its charge center therefore is located significantly closer to the nucleus than the centers of bonding electron pairs. Lone electron pairs are especially effective sterically in the following manner:

- If the polyhedron has non-equivalent vertices, a lone electron pair will take the position that is most distant from the remaining electron pairs. At a trigonal bipyramid these are the equatorial positions. In SF_4 and in ClF_3 the lone electron pairs thus take equatorial positions, and fluorine atoms take the two axial positions (cf. Table 8).

- Two equivalent lone electron pairs take those positions which are farthest apart. In an octahedron two lone electron pairs thus have the *trans* configuration, as for example in XeF_4.

- Due to their stronger repulsion lone electron pairs press other electron pairs closer together. The more lone electron pairs that are present, the more this effect is noticeable, and the more the real molecular geometry deviates from the ideal polyhedron. For larger central atoms the charge centers of bonding electron pairs are more distant from each other, their mutual repulsion is reduced and the lone electron pairs can press them together to a larger extent. The following bond angles exemplify this:

Table 8 Molecular structures of compounds AX_nE_m. A = main group element, E = lone electron pair

composition	structure	angle XAX	examples
AX_2E		$< 120°$	$SnCl_2(g)$, $GeBr_2(g)$
AX_2E_2		$< 109.5°$	H_2O, F_2O, Cl_2O, H_2S, H_2N^-
AX_2E_3		$180°$	XeF_2, ICl_2^-, I_3^-
AX_3E		$< 109.5°$	NH_3, NF_3, PH_3, PCl_3, OH_3^+, SCl_3^+, $SnCl_3^-$
AX_3E_2		$< 90°$	ClF_3
AX_4E		$< 90°$ and $< 120°$	SF_4
AX_4E_2		$90°$	XeF_4, BrF_4^-, ICl_4^-
AX_5E		$< 90°$	$SbCl_5^{2-}$ SF_5^-, BrF_5
AX_5E_2		$72°$	XeF_5^-

CH$_4$		NH$_3$		OH$_2$
109.5°	>	107.3°	>	104.5°
		V		V
SiH$_4$		PH$_3$		SH$_2$
109.5°	>	93.5°	>	92.3°
		V		V
GeH$_4$		AsH$_3$		SeH$_2$
109.5°	>	92.0°	>	91.0°
		V		V
SnH$_4$		SbH$_3$		TeH$_2$
109.5°	>	91.5°	>	89.5°

If only an unpaired electron is present instead of an electron pair, its influence is reduced, for example:

$$\langle O{=}N{=}O\rangle^{+} \qquad {}'O{=}\dot{N}{=}O\backslash \qquad {}|O{\overset{\displaystyle\bar{N}}{\cdots}}O|\;^{-}$$

$$180° \qquad\qquad 134° \qquad\qquad 115°$$

By definition, the coordination number includes the adjacent atoms, and lone electron pairs are not counted. On the other hand, now we are considering lone pairs as occupying polyhedron vertices. To take account of this, we regard the coordination sphere as including the lone pairs, but we designate them with a ψ, for example: ψ_2-octahedral = octahedron with two lone electron pairs and four ligands.

2. Decreasing **electronegativity of the ligand atoms** causes the charge centers of bonding electron pairs to shift toward the central atom; their repulsive activity increases. Ligands having low electronegativity therefore have a similar influence, albeit less effective, as lone electron pairs. Correspondingly the bond angles in the following pairs increase:

F$_2$O		H$_2$O		NF$_3$		NH$_3$
103.2°	<	104.5°		102.1°	<	107.3°

However, bond angles cannot be understood satisfactorily by considering only the electron-pair repulsion because they also depend on another factor:

3. The effective size of the ligands. In most cases (unless the central atom is very large) the ligands come closer to each other than the corresponding VAN DER WAALS distance; the electron shells of the ligand atoms interpenetrate one another and additional repulsive forces become active. This is especially valid for large ligand atoms. Within a group of the periodic table decreasing electronegativities and increasing atom sizes go hand in hand, so that they act in the same sense. The increase of the following bond angles is due to both effects:

	HCF_3		$HCCl_3$		$HCBr_3$		HCl_3
Hal–C–Hal	108.8°	<	110.4°	<	110.8°	<	113.0°

	PF_3		PCl_3		PBr_3		PI_3
Hal–P–Hal	97.8°	<	100.1°	<	101.0°	<	102°

When the electronegativity and the size of the ligand atoms have opposing influence, no safe predictions can be made:

F_2O H_2O influence of the
103.2°   <   104.5°   electronegativity predominates

Cl_2O H_2O influence of the
110.8°   >   104.5°   ligand size predominates

Sometimes the opposing influence of the two effects is just balanced. For example, the sterical influence of chlorine atoms and methyl groups is often the same (the carbon atom of a methyl group is smaller, but it is less electronegative than a chlorine atom):

Cl_2O Me_2O PCl_3 PMe_3
110.8°   111°   100.1°   99.1°

4. A pre-existing distortion is found when certain bond angles deviate from the ideal values of the corresponding polyhedron for geometrical reasons. In this case the remaining angles adapt themselves. Forced angle deviations result mainly in small rings. For example:

The bridging chlorine atoms (2 lone electron pairs) should have bond angles smaller than 109.5° but larger than 90°. The angle at the metal atom in the four-membered ring is forced to a value under 90°; it adopts 78.6°. The outer, equatorial Cl atoms now experience a reduced repulsion, so that the angle between them is enlarged from 90° to 101.2°. Due to this distortion the axial Cl atoms should be inclined slightly outwards; however, as the Nb–Cl bonds in the ring are longer and their charge centers therefore are situated farther away from the centers of the niobium atoms, they are less repulsive, and the axial Cl atoms are inclined inwards. The increased Nb–Cl distances in the ring are a consequence of the higher coordination number (2 instead of 1) at the bridging Cl atoms (cf. point 6, p. 55).

5. Multiple bonds can be treated as ring structures with bent bonds. The distortions dealt with in the preceding paragraph must be taken into account. For instance, in ethylene every C atom is surrounded tetra-hedrally by four electron pairs; two pairs mediate the double bond be-tween the C atoms via two bent bonds. The tension in the bent bonds reduces the angle between them and decreases their repulsion toward the C–H bonds, and the HCH bond angle is therefore bigger than 109.5°.

Usually, it is more straightforward to treat double and triple bonds as if they form a single orbital that is occupied by four and six electrons, respectively. The increased repulsive power of this orbital corresponds to its high charge. In this way, the structure of the ethylene molecule can be considered as having triangularly coordinated C atoms, but with angles deviating from 120°; the two angles between the double bond and the C–H bonds will be larger than 120°, and the H–C–H will be smaller than 120°. Similarly, a molecule like $OPCl_3$ will have a tetrahedral structure, but with widened OPCl bonds. When two double bonds are present, the angle between them will be the largest. In Table 9 some examples are listed; they also show the influence of electronegativity and atomic size.

Table 9 Bond angles in degrees for some molecules having multiple bonds. X = singly, Z = double bonded ligand atom

		XAX		XAX		XAX	ZAZ		XAX	XAZ
$F_2C=O$	107.7	$F_3P=O$	102.5	F_2SO_2	98.6	124.6	$F_2S=O$	92.2	106.2
$Cl_2C=O$	111.3	$Cl_3P=O$	103.6	Cl_2SO_2	101.8	122.4	$Cl_2S=O$	97.0	108.0
$Me_2C=O$	109	$Br_3P=O$	104				$Br_2S=O$	96	108
$H_2C=O$	120.4	$Me_3P=O$	104.1	Me_2SO_2	115	125	$Me_2S=O$	100	109

6. Bond lengths are affected like bond angles. The more electron pairs are present, the more they repel each other and the longer the bonds become. The increase in bond length with increasing coordination number is also mentioned in the discussion of ionic radii (p. 34). For example:

distance Sn–Cl: $SnCl_4$ 228 pm $Cl_4Sn(OPCl_3)_2$ 233 pm

The polarity of bonds indeed has a much more marked influence on their lengths. With increasing negative charge of a particle the repulsive forces gain importance. Examples:

distance Sn–Cl: $Cl_4Sn(OPCl_3)_2$ 233 pm $SnCl_6^{2-}$ 244 pm

	P–O /pm	P–F /pm	O–P–O /°	F–P–F /°
POF_3	144	152	–	101.3
$PO_2F_2^-$	147	157	122	97
PO_3F^{2-}	151	159	114	–
PO_4^{3-}	155	–	109.5	–

The bond between two atoms of different electronegativity is polar. The opposite partial charges of the atoms cause them to attract each other. A change in polarity affects the bond length. This can be noted especially when the more electronegative atom participates in more bonds than it should have according to the $8 - N$ rule: contrary to its electronegativity it must supply electrons for the bonds, its negative partial charge is lowered or even becomes positive, and the attraction to the partner atom is decreased. This effect is conspicuous for bridging halogen atoms, as can be seen by comparison with the non-bridging atoms:

Niobium pentachloride mentioned on p. 54 is another example. The BAUR rules also express these facts (cf. p. 46).

7. Influence of a partial valence shell. Atoms of elements of the third period, such as Si, P, and S, and of higher periods can accomodate more than four valence electron pairs in their valence shell. By resorting to d states an accomodation of up to nine electron pairs is conceivable. As a matter of fact, most main group elements only tolerate a maximum of six electron pairs (for example the S atom in SF_6). Compounds having more than six valence electron pairs per atom are known only for the heavy elements, the iodine atom in IF_7 being an example. Obviously an increased repulsion between the electron pairs comes into effect when the bond angles become smaller than 90°; this would have to be the case for coordination numbers higher than six. However, crowding electron pairs down to angles of 90° is possible without large resistance (cf. hydrogen compounds listed on p. 53; note the marked jump for the angles between the second and the third period).

If the central atom still can take over electrons and if a ligand disposes of lone electron pairs, then these tend to pass over to the central atom to some degree. In other words, the electron pairs of the ligand reduce their mutual repulsion by shifting partially towards the central atom. This applies especially for small ligand atoms like O and N, particularly when high formal charges have to be allocated to them. For this reason terminal O and N atoms tend to form multiple bonds with the central atom, for example:

A similar explanation can be given for the larger Si–O–Si bond angles as compared to C–O–C. Electron density is given over from the oxygen atom into the valence shells of the silicon atoms, but not of the carbon atoms, in the sense of the resonance formulas:

$$Si-\overline{\underline{O}}-Si \quad \longleftrightarrow \quad Si=O=Si \qquad C^{\overset{\frown}{\overline{O}}}C$$

Examples:

angle SiOSi		angle COC	
O(SiH$_3$)$_2$	144°	O(CH$_3$)$_2$	111°
α-quartz	142°	O(C$_6$H$_5$)$_2$	124°
α-cristobalite	147°		

That the COC angle is larger in diphenyl ether than in diethyl ether can be explained by the electron acceptor capacity of the phenyl groups. When two strongly electron-accepting atoms are bonded to an oxygen atom, the transition of the lone electron pairs can go so far that a completely linear group of atoms M=O=M results, as for example in the [Cl$_3$FeOFeCl$_3$]$^{2-}$ ion.

The electron transition and the resulting multiple bonds should reveal themselves by shortened bond lengths. In fact, bridging oxygen atoms between metal atoms in a linear arrangement exhibit rather short metal–oxygen bonds. Because of the high electronegativity of oxygen the charge centers of the bonding electron pairs will be located more towards the side of the oxygen atom, i.e. the bonds will be polar. This can be expressed by the following resonance formulas:

$$[\ Cl_3 \overset{2\ominus}{Fe}=\overset{2\oplus}{O}=\overset{2\ominus}{Fe}Cl_3 \quad \longleftrightarrow \quad Cl_3Fe\ |\overset{2\ominus}{\underline{O}}|\ FeCl_3\ ]^{2-}$$

The lowest formal charges result when both formulas have equal weight. In the case of bridging fluorine atoms the ionic formula should be more important to achieve lower formal charges, e.g.:

$$[\ F_5 \overset{2\ominus}{Sb}=\overset{3\oplus}{F}=\overset{2\ominus}{Sb}F_5 \quad \longleftrightarrow \quad F_5Sb\ |\overset{\ominus}{\underline{F}}|\ SbF_5\ ]^{-}$$

As a matter of fact, bridging fluorine atoms usually exhibit bond angles between 140 and 180° and rather long bonds. The significance of the right formula, representing weak interactions between the central F$^-$ ion and two SbF$_5$ molecules, also shows up in the chemical reactivity: fluorine bridges are cleaved easily.

Restrictions

The consideration of the mutual valence electron-pair repulsion as a rule results in correct qualitative models for molecular structures. In spite of the simple concept the theory is well-founded and compatible with the more complicated and less illustrative MO theory (chapter 9). The results often are by no means inferior to those of sophisticated calculations. Nevertheless, in some cases the model fails. Examples are the ions $SbBr_6^{3-}$, $SeBr_6^{2-}$, and $TeCl_6^{2-}$ which have undistorted octahedral structures, although the central atom still has a lone electron pair. This electron pair is said to be "stereochemically inactive", although this is not quite true, because its influence still shows up in increased bond lengths. This phenomenon is observed only for higher coordination numbers (≥ 6), when the central atom is a heavy atom and when the ligands belong to a higher period of the periodic table, i.e. when the ligands are easily polarized. The decreasing influence of the lone electron pairs can also be seen by comparing the solid state structures of AsI_3, SbI_3 and BiI_3. AsI_3 forms pyramidal molecules (bond angles $100.2°$), but in the solid they are associated in that three iodine atoms of adjacent molecules are coordinated to an arsenic atom. In all, the coordination is distorted octahedral, with three intramolecular As–I lengths of 259 pm and three intermolecular lengths of 347 pm. In BiI_3 the coordination is octahedral with six equal Bi–I distances (307 pm). SbI_3 takes an intermediate position ($3\times$ 287 pm, $3\times$ 332 pm).

The theory also cannot explain the "*trans*-influence" that is observed between ligands that are located on opposite sides of the central atom on a straight line, as for two ligands in *trans* arrangement in octahedral coordination. The more tightly one ligand is bonded to the central atom, as evidenced by a short bond length, the longer is the bond to the ligand in the *trans* position. Particularly multiple bonds are strongly effective in this way. This can also be noted in the reactivity: the weakly bonded ligand is easily displaced. On the other hand, a lone electron pair usually does not cause a lengthened bond of the ligand *trans* to it; on the contrary, this bond tends to be shorter (distances in pm):

Note how the bond angles show the repulsive action of the multiple bonds and of the lone electron pair, respectively.

In one respect the valence shell electron-pair repulsion theory is no better (and no worse) than other theories of molecular structure. Predictions can only be made when the constitution is known, i.e. when it is already known which and how many atoms are joined to each other. For instance, it cannot be explained why the following pentahalides consist of so different kinds of molecules or ions in the solid state: $SbCl_5$ monomer, $(NbCl_5)_2$ dimer, $(PaCl_5)_\infty$ polymer, $PCl_4^+PCl_6^-$ ionic, $PBr_4^+Br^-$ ionic; PCl_2F_3 monomer, $AsCl_4^+AsF_6^-$ ($= AsCl_2F_3$) ionic, $SbCl_4^+[F_4ClSb-F-SbClF_4]^-$ ($= SbCl_2F_3$) ionic.

7.2   Structures with Five Valence Electron Pairs

The features described in the preceding section can be studied well with molecules for which five valence electron pairs are to be considered. As they also show some peculiarities, we will deal with them separately. The favored arrangement for five points on a sphere is the trigonal bipyramid. Its two axial and three equatorial positions are not equivalent, a greater repulsive force being excercised on the axial positions. Therefore, lone electron pairs as well as ligands with lower electronegativities prefer the equatorial sites. If the five ligands are of the same kind, the bonds to the axial ligands are longer (in other words, the covalent radius is larger in the axial direction). Cf. Table 10.

The molecular parameters for CH_3PF_4 and $(CH_3)_2PF_3$ illustrate the influence of the lower electronegativity of the methyl groups and the corresponding increased repulsive effect of the P–C bond electron pairs:

Energetically, the tetragonal pyramid is almost as favorable as the trigonal bipyramid. With a bond angle of 104° between the apical and a basal position the repulsion energy is only 0.14 % more when a purely coulombic repulsion is assumed; for "hard" orbitals the difference is even less. Furthermore, the transformation of a trigonal bipyramid to a tetragonal pyramid requires only a low activation energy; as a consequence, a fast exchange of positions of the ligands from one trigonal bipyramid to a tetragonal pyramid and on to a differently oriented trigonal bipyramid occurs ("BERRY rotation", Fig. 22). This explains why only a doublet peak is observed in the ^{19}F-NMR-spectrum of PF_5 even at low temperatures; the doublet is due to the P–F spin-spin coupling. If the fast positional exchange did not occur, two doublets with an intensity ratio of 2:3 would be expected.

The tetragonal pyramid ususally is favored when one double bond is present. Molecules or ions such as $O=CrF_4$, $O=WCl_4$ (as monomers in the gas phase), $O=TiCl_4^{2-}$ or $S=NbCl_4^-$ have this structure. However, $O=SF_4$ has a trigonal-bipyramidal structure with the oxygen atom in an equatorial positions.

Table 10 Axial and equatorial bond lengths (/pm) for trigonal-bipyramidal distribution of the valence elecrons

AX_5	AX_{ax}	AX_{eq}	AX_4E	AX_{ax}	AX_{eq}	AX_3E_2	AX_{ax}	AX_{eq}
PF_5	158	153	SF_4	165	155	ClF_3	170	160
AsF_5	171	166	SeF_4	177	168	BrF_3	181	172
PCl_5	212	202						

Very low energy differences also result for different polyhedra with higher coordination numbers, including coordination number 7. In these cases the electron pair repulsion theory no longer allows reliable predictions.

Fig. 22
Change of ligand positions between trigonal bipyramids and a tetragonal pyramid

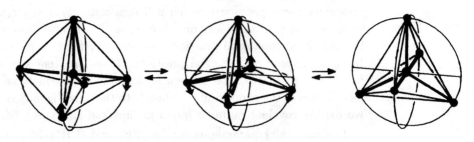

7.3 Problems

7.1 What structures will the following molecules have according to VSEPR theory?
$BeCl_2(g)$, BF_3, PF_3, BrF_3, $TeCl_3^+$, XeF_3^+, $GeBr_4$, $AsCl_4^+$, SbF_4^-, ICl_4^-, BrF_4^+, $TiBr_4$, $SbCl_5$, $SnCl_5^-$, TeF_5^-, $ClSF_5$, O_3^-, Cl_3^-, S_3^{2-}, O_2ClF_3, $O_2ClF_2^-$, $OClF_4^-$, O_3BrF, O_3XeF_2.

7.2 The following dimer species are associated via two bridging chloro atoms. What are their structures?
Be_2Cl_4, Al_2Br_6, I_2Cl_6, $As_2Cl_8^{2-}$, Ta_2I_{10}.

7.3 What structure is to be expected for $H_2C=SF_4$?

7.4 Arrange the following molecules in the order of increasing bond angles.
(a) OF_2, SF_2, SCl_2, S_3^-, S_3^{2-};
(b) Angle H–N–H in H_3CNH_2, $[(H_3C)_2NH_2]^+$;
(c) Angle F_{ax}–P–F_{ax} in PCl_2F_3, PCl_3F_2.

7.5 There are two bridging chloro atoms in Al_2Cl_6. Give the sequence of increasing bond lengths and bond angles and estimate approximate values for the angles.

7.6 Which of the following species should have the longer bond lengths?
$SnCl_3^-$ or $SnCl_5^-$; PF_5 or PF_6^-; $SnCl_6^{2-}$ or $SbCl_6^-$.

7.7 Which of the following species are the most likely to violate the VSEPR rules?
SbF_5^{2-}, $BiBr_5^{2-}$, TeI_6^{2-}, ClF_5, IF_7, IF_8^-.

8 Molecular Structures II: Compounds of Transition Metals

8.1 Ligand Field Theory

The mutual interaction between *bonding* electron pairs is the same for transition metal compounds as for compounds of main group elements. All statements concerning molecular structure apply equally. However, *nonbonding* valence electrons behave differently. For transition metal atoms these generally are d electrons that can be accomodated in five d orbitals. In what manner the electrons are distributed among these orbitals and in what way they become active stereochemically can be judged with the aid of *ligand field theory*. The concept of ligand field theory is equivalent to that of the valence shell electron-pair repulsion theory: it considers how the d electrons have to be distributed so that they attain a minimum repulsion with each other *and* with the bonding electron pairs. In its original version by H. BETHE it was formulated as crystal field theory; it considered the electrostatic repulsion between the d electrons and the ligands, which were treated as point-like ions.* After the success of the valence shell electron-pair repulsion theory it appears more appropriate to consider the interactions between nonbonding d electrons and bonding electron pairs; thus the same concepts apply for both theories. This way, one obtains qualitatively correct structural statements with relatively simple models. The more exact molecular orbital theory draws the same conclusions.

The relative orientations of the regions with high charge density of d electrons and of bonding electrons about an atom can be described with the aid of a coordinate system that has its origin in the center of the atom. Two sets of d orbitals are to be distinguished (Fig. 23): the first set consists of two orbitals oriented along the coordinate axes, and the second set consists of three orbitals oriented toward the centers of the edges of an circumscribed cube.

Octahedral Coordination

If an atom has six ligands, then the mutual repulsion of the six bonding electron pairs results in an octahedral coordination. The positions of the ligands correspond to points on the axes of the coordinate system. If nonbonding electrons are present, these will prefer the orbitals d_{xy}, d_{yz} and d_{xz} because the regions of high charge density of the other two d orbitals are especially close to the

---

*The terms crystal field theory and ligand field theory are not used in a uniform way. As only interactions between adjacent atoms are being considered, without referring to crystal influences, the term crystal field theory does not seem adequate. Some authors consider certain electronic interactions (like π bonds) as part of ligand field theory, although they originate from MO theory.

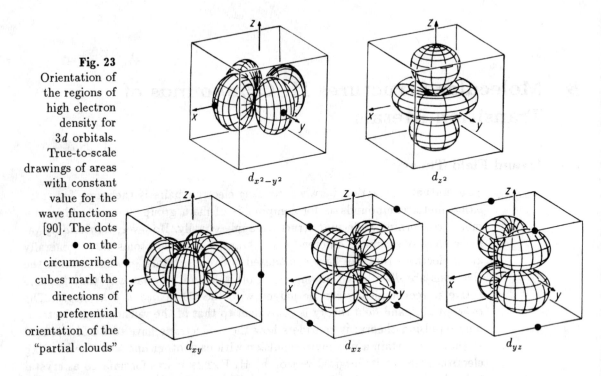

Fig. 23
Orientation of the regions of high electron density for $3d$ orbitals. True-to-scale drawings of areas with constant value for the wave functions [90]. The dots • on the circumscribed cubes mark the directions of preferential orientation of the "partial clouds"

$d_{x^2-y^2}$ d_{z^2}

d_{xy} d_{xz} d_{yz}

bonding electron pairs. The three orbitals favored energetically are termed t_{2g} orbitals (this is a symbol for the orbital symmetry; the t designates a triply degenerate state); the other two are e_g orbitals (e = doubly degenerate; from German *entartet* = degenerate). Cf. the diagram in the margin.

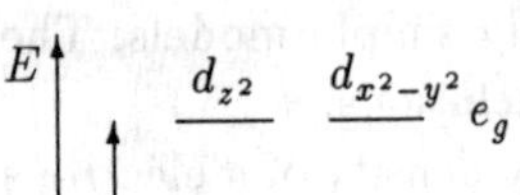

The energy difference between the occupation of a t_{2g} and an e_g orbital is termed Δ_o. The value of Δ_o depends on the repulsion exercised by the bonding electron pairs on the d electrons. Compared to a transition metal atom the bonded ligand atoms are usually much more electronegative. The centers of charge of the bonding electron pairs are much closer to them, especially when they are strongly electronegative. Therefore, one can expect a decreasing influence on the d electrons and thus a decrease of Δ_o with increasing ligand electronegativity. Decreasing Δ_o values also result with increasing sizes of the ligand atoms; in this case the electron pairs are distributed over a larger space so that the difference of their repulsive action on a t_{2g} and an e_g orbital is less marked. In the presence of multiple bonds between the metal atom and the ligands, as for example in metal carbonyls, the electron denstity of the bonds is especially high and their action is correspondingly large. Since Δ_o is a value that can be measured directly with spectroscopic methods, the activities of different kinds of ligands are well-known.* The *spectrochemical series* is obtained by ordering different ligands according to decreasing Δ_o:

Margin diagram:

E

d_{z^2} $d_{x^2-y^2}$ e_g

Δ_o

d_{xy} d_{xz} d_{yz} t_{2g}

---

*By photoexcitation of an electron from the t_{2g} to the e_g level we have $\Delta_o = h\nu$

$$CO > CN^- > PR_3 > NO_2^- > NH_3 > NCS^- > H_2O > RCO_2^- \approx OH^-$$
$$> F^- > NO_3^- > Cl^- \approx SCN^- > S^{2-} > Br^- > I^-$$

When two or three nonbonding electrons are present, they will occupy two or three of the t_{2g} orbitals, respectively (HUND's rule). This is more favorable than pairing electrons in one orbital because the pairing requires that the electrostatic repulsion between the two electrons be overcome. The energy necessary to include a second electron in an already occupied orbital is called the electron pairing energy P. When four nonbonding electrons are present, there are two alternatives for the placement of the fourth electron. If $P > \Delta_o$, then it will be an e_g orbital and all four electrons will have parallel spin: we call this a *high-spin complex*. If $P < \Delta_o$, then it is more favorable to form a *low-spin complex* leaving the e_g orbitals unoccupied and having two paired electrons:

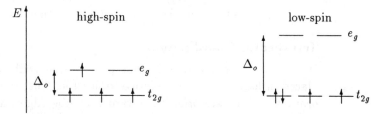

In a high-spin d^4 complex only one of the two e_g orbitals is occupied. If it is the d_{z^2} orbital then it exerts a strong repulsion on the bonding electrons of the two ligands on the z axis. These ligands are forced outwards; the coordination octahedron suffers an elongation along the z axis. This effect is known as the *Jahn–Teller effect*. Instead of the d_{z^2} orbital the $d_{x^2-y^2}$ orbital could have been occupied, which would have produced elongations along the x and y axes. However, a higher force is needed to stretch four bonds; stretching only two bonds is energetically more favorable, and consequently only examples with octahedra elongated in one direction are known so far.[*]

The JAHN–TELLER effect is always to be expected when degenerate orbitals are unevenly occupied with electrons. As a matter of fact, it is observed for the following electronic configurations:

	d^4 high-spin	d^9	d^7 low-spin
Examples	Cr(II), Mn(III)	Cu(II)	Ni(III)

A JAHN–TELLER distortion should also occur for configuration d^1. However, in this case the occupied orbital is a t_{2g} orbital, for example d_{xy}; this exerts a repulsion on the ligands on the axes x and y which is only slightly larger than the force exerted along the z axis. The distorting force is usually not sufficient

---

[*]Contrary examples in the literature have turned out to be erroneous

to produce a perceptible effect. Ions like TiF_6^{3-} or $MoCl_6^-$, for example, show no detectable deviation from octahedral symmetry.

Not even the slightest JAHN–TELLER distortion and therefore no deviation from the ideal octahedral symmetry is to be expected when the t_{2g} and e_g orbitals are occupied evenly. This applies for the following electronic configurations:

d^0, d^3, d^5 high-spin, d^6 low-spin, d^8 and d^{10}. For configuration d^8, octahedral coordination occurs only seldomly, however (see below, square coordination).

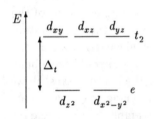

If there are different kinds of ligands, those which have the smaller influence according to the spectrochemical series prefer the positions with the stretched bonds. For example, in the $[CuCl_4(OH_2)_2]^{2-}$ ion two of the Cl atoms take the positions in the vertices of the elongated axis of the coordination polyhedron.

Tetrahedral Coordination

We can imagine the four ligands of a tetrahedrally coordinated atom to be placed in four of the eight vertices of a cube. The orbitals d_{xy}, d_{yz} and d_{xz} (t_2 orbitals), which are oriented toward the cube edges, are closer to the bonding electron pairs than the orbitals $d_{x^2-y^2}$ and d_{z^2} (e orbitals). Consequently, the t_2 orbitals experience a larger repulsion and become energetically higher than the e orbitals; the sequence is opposite to that of octahedral coordination. The energy difference is termed Δ_t. Since none of the d orbitals is oriented toward a cube vertex, $\Delta_t < \Delta_o$ is expected (for equal ligands, equal central atom and equal bond lengths), or, more specifically $\Delta_t \approx \frac{4}{9}\Delta_o$. Δ_t always is smaller than the spin pairing energy; tetrahedral complexes always are high-spin complexes.

If the t_2 orbitals are occupied unevenly, JAHN–TELLER distortions occur. For configuration d^4, one of the t_2 orbitals is unoccupied; for d^9, one has single occupation and the rest double. As a consequence, the ligands experience differing repulsions, and a flattened tetrahedron results (Fig. 24). Typical bond angles are, for example in the $CuCl_4^{2-}$ ion, $2 \times 116°$ and $4 \times 106°$.

For the configurations d^3 and d^8 one t_2 orbital has one electron more than the others; in this case an elongated tetrahedron is to be expected; however, the deformation turns out to be smaller than for d^4 and d^9, because the deforming repulsion force is being exerted by only one electron (instead of two; Fig. 24). Since the deformation force is small and the requirements of the packing in the crystal sometimes cause opposite deformations, observations do not always conform to expectations. For example, $NiCl_4^{2-}$ (d^8) has been observed to have undistorted, slightly elongated or slightly flattened tetrahedra depending on the cation. For uneven occupation of the e orbitals distortions could also be expected, but the effect is even smaller and usually it is not detectable; VCl_4 (d^1) for example has undistorted tetrahedra.

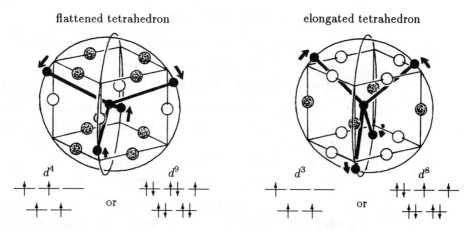

flattened tetrahedron

elongated tetrahedron

d^4 d^9 d^3 d^8

Fig. 24 JAHN–TELLER distortions of tetrahedral complexes. The arrows indicate the directions of displacement of the ligands due to repulsion by the nonbonding d electrons. The spheres on the cube edges mark the centers of gravity of the charges of the t_2 orbitals, a hatched sphere being occupied by one electron more than a white sphere

Square Coordination

When the two ligands on the z axis of an octahedral complex are removed, the remaining ligands form a square. The repulsion between bonding electrons on the z axis ceases for the d_{z^2}, the d_{xz}, and the d_{yz} electrons. Only one orbital, namely $d_{x^2-y^2}$, still experiences a strong repulsion from the remaining bond electrons and is energetically unfavorable (Fig. 25). Square coordination is the preferential coordination for d^8 configuration, as for Ni(II) and especially for Pd(II), Pt(II), and Au(III), in particular with ligands that cause a strong splitting of the energy levels. Both an octahedral complex (two electrons in e_g orbitals) and a tetrahedral complex (four electrons in t_2 orbitals) are less favorable in this case.

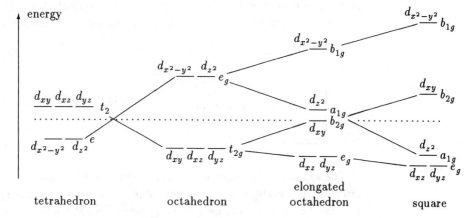

Fig. 25 Diagram of the relative energies of electrons in d orbitals for different geometrical arrangements. The "centers of gravity" (mean values of the energy levels) for all term sequences were positioned on the dotted line

8.2 Ligand Field Stabilization Energy

When ligands approach a central atom or ion, the following energetical contributions become effective:

- Energy gain (freed energy) by the formation of covalent bonds.

- Energy expenditure due to the mutual repulsion of the bonding electron pairs and due to the repulsion between ligands that approach each other too closely.

- Energy expenditure due to the repulsion exerted by bonding electron pairs on nonbonding electrons of the central atom.

Ligand field theory mainly considers the last contribution. For this contribution the geometrical distribution of the ligands is irrelevant as long as the electrons of the central atom have a spherical distribution; the repulsion energy always is the same in this case. All half and fully occupied electron shells of an atom are spherical, including d^5 high-spin and d^{10} (and naturally d^0). This is not so for other d electron configurations.

In order to compare the structural options for transition metal compounds and to estimate which of them are most favorable energetically, the *ligand field stabilization energy* (LFSE) is a useful parameter. This is defined to be the difference between the repulsion energy of the bonding electrons toward the d electrons as compared to a notional repulsion energy that would exist if the d electron distribution were spherical.

In an octahedral complex a d_{z^2} electron is oriented toward the ligands (the same applies for $d_{x^2-y^2}$); it exercises more repulsion than if it were distributed spherically. Compared to this imaginary distribution it has a higher energy state. On the other hand, a d_{xy} electron is lowered energetically: it is being repelled less than an electron with spherical distribution. The *principle of the weighted mean* holds: the sum of the energies of the raised and the lowered states must be equal to the energy of the spherical state. Since there are two raised and three lowered states for an octahedron, the following scheme results:

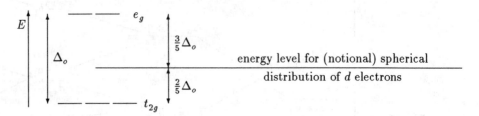

The energy level diagrams in Fig. 25 have been drawn according to the principle of the weighted mean energy. They show how the energy levels are placed relative to the level of the notional state of a spherical d electron distribution. They do not represent absolute energy values, as the absolute level of the notional state also depends on the other energy contributions mentioned above.

Table 11 Ligand field stabilization energies (LFSE) for octahedral and tetrahedral ligand distributions

	0	1	2	3	4	5	6	7	8	9	10
				number of d electrons							
octahedra, high-spin				electron distribution × energy $/\Delta_o$							
$\frac{3}{5}\Delta_o$ ... e_g	0	0	0	0	$1 \cdot \frac{3}{5}$	$2 \cdot \frac{3}{5}$	$2 \cdot \frac{3}{5}$	$2 \cdot \frac{3}{5}$	$2 \cdot \frac{3}{5}$	$3 \cdot \frac{3}{5}$	$4 \cdot \frac{3}{5}$
$-\frac{2}{5}\Delta_o$ ... t_{2g}	0	$-1 \cdot \frac{2}{5}$	$-2 \cdot \frac{2}{5}$	$-3 \cdot \frac{2}{5}$	$-3 \cdot \frac{2}{5}$	$-3 \cdot \frac{2}{5}$	$-4 \cdot \frac{2}{5}$	$-5 \cdot \frac{2}{5}$	$-6 \cdot \frac{2}{5}$	$-6 \cdot \frac{2}{5}$	$-6 \cdot \frac{2}{5}$
sum = LFSE $/\Delta_o$	0	$-\frac{2}{5}$	$-\frac{4}{5}$	$-\frac{6}{5}$	$-\frac{3}{5}$	0	$-\frac{2}{5}$	$-\frac{4}{5}$	$-\frac{6}{5}$	$-\frac{3}{5}$	0
octahedra, low-spin				electron distribution × energy $/\Delta_o$							
$\frac{3}{5}\Delta_o$ ... e_g	0	0	0	0	0	0	0	$1 \cdot \frac{3}{5}$	$2 \cdot \frac{3}{5}$	$3 \cdot \frac{3}{5}$	$4 \cdot \frac{3}{5}$
$-\frac{2}{5}\Delta_o$ ... t_{2g}	0	$-1 \cdot \frac{2}{5}$	$-2 \cdot \frac{2}{5}$	$-3 \cdot \frac{2}{5}$	$-4 \cdot \frac{2}{5}$	$-5 \cdot \frac{2}{5}$	$-6 \cdot \frac{2}{5}$	$-6 \cdot \frac{2}{5}$	$-6 \cdot \frac{2}{5}$	$-6 \cdot \frac{2}{5}$	$-6 \cdot \frac{2}{5}$
sum = LFSE $/\Delta_o$	0	$-\frac{2}{5}$	$-\frac{4}{5}$	$-\frac{6}{5}$	$-\frac{8}{5}$	$-\frac{10}{5}$	$-\frac{12}{5}$	$-\frac{9}{5}$	$-\frac{6}{5}$	$-\frac{3}{5}$	0
tetrahedra, high-spin				electron distribution × energy $/\Delta_t$							
$\frac{2}{5}\Delta_t$ ... t_2	0	0	0	$1 \cdot \frac{2}{5}$	$2 \cdot \frac{2}{5}$	$3 \cdot \frac{2}{5}$	$3 \cdot \frac{2}{5}$	$3 \cdot \frac{2}{5}$	$4 \cdot \frac{2}{5}$	$5 \cdot \frac{2}{5}$	$6 \cdot \frac{2}{5}$
$-\frac{3}{5}\Delta_t$ ... e	0	$-1 \cdot \frac{3}{5}$	$-2 \cdot \frac{3}{5}$	$-2 \cdot \frac{3}{5}$	$-2 \cdot \frac{3}{5}$	$-2 \cdot \frac{3}{5}$	$-3 \cdot \frac{3}{5}$	$-4 \cdot \frac{3}{5}$	$-4 \cdot \frac{3}{5}$	$-4 \cdot \frac{3}{5}$	$-4 \cdot \frac{3}{5}$
sum = LFSE $/\Delta_t$	0	$-\frac{3}{5}$	$-\frac{6}{5}$	$-\frac{4}{5}$	$-\frac{2}{5}$	0	$-\frac{3}{5}$	$-\frac{6}{5}$	$-\frac{4}{5}$	$-\frac{2}{5}$	0

Even when the central atoms and the ligands are the same, the level of the notional state differs on an absolute scale for different ligand arrangements, i.e. the different term schemes are shifted mutually.

Table 11 lists the values for the ligand field stabilization energies for octahedral and tetrahedral complexes. The values are given as multiples of Δ_o and Δ_t, respectively. In Fig. 26 the values have been plotted; the curves also show the influence of the other energy contributions for $3d$ elements. In the series from Ca^{2+} to Zn^{2+} the ionic radii decrease and the bond energies increase; correspondingly the curves run downwards from left to right. The dashed lines apply for the notional ions with spherical electron distributions; the actual energy values for the truly spherical electron distributions d^0, d^5 high-spin and d^{10} are situated on these lines. Due to the decreasing ionic radii octahedral complexes become less stable than tetrahedral complexes toward the end of the series (because of increasing repulsive forces between the bonding electron pairs and due to the more crowded ligand atoms); for this reason the dashed line for octahedra bends upwards at the end. The ligand field stabilization energy is the reason for the occurence of two minima in the curves for high-spin complexes; the minima correspond to the configurations d^3 and d^8 for octahedral and to d^2 and d^7 for tetrahedral complexes. The stabilization energies are less for tetrahedral ligand fields, since generally $\Delta_o > \Delta_t$ (in Fig. 26 $\Delta_t = \frac{4}{9}\Delta_o$ was assumed). For octahedral low-spin complexes there is only one minimum at d^6.

For high-spin compounds only rather small stabilization differences result between octahedral and tetrahedral coordination for the configurations d^7 and d^8 (Fig. 26). Co^{2+} shows a tendency to tetrahedral coordination, whereas this

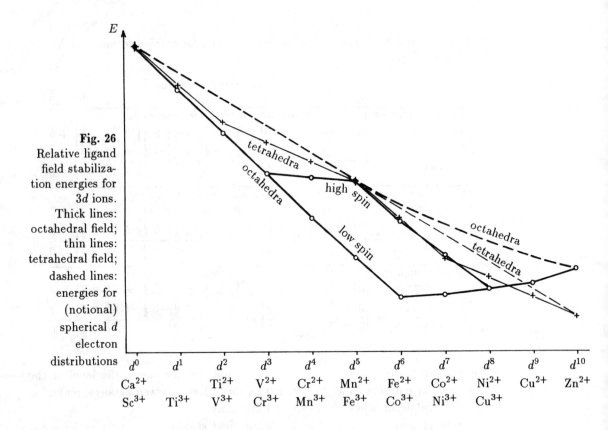

Fig. 26
Relative ligand field stabilization energies for $3d$ ions. Thick lines: octahedral field; thin lines: tetrahedral field; dashed lines: energies for (notional) spherical d electron distributions

tendency is overcompensated for Ni^{2+} by the larger ligand field stabilization for octahedra, so that Ni^{2+} prefers octahedral coordination. Here the different locations of the maxima of the ligand field stabilization energies takes effect (Table 11): it is largest for tetrahedra at configuration d^7 (Co^{2+}) and for octahedra at d^8 (Ni^{2+}). With increasing ligand sizes the tendency toward tetrahedral coordination becomes more marked; in other words, the octahedral arrangement becomes relatively less stable; in Fig. 26 this would be expressed by an earlier upwards bending of the thick dashed line. Fe^{2+} and Mn^{2+} also form tetrahedral complexes with larger ligands like Cl^- or Br^-.

In Fig. 26 the additional stabilization by the JAHN–TELLER effect has not been taken into account. Its inclusion brings the point for the (distorted) octahedral coordination for Cu^{2+} further down, thus rendering this arrangement more favorable.

The ligand field stabilization is expressed in the lattice energies of the halides MX_2. The values obtained by the BORN–HABER cycle from experimental data are plotted vs. the d electron configuration in Fig. 27. The ligand field stabilization energy contribution is no more than 200 kJ mol^{-1}, which is less than 8% of the total lattice energy. The ionic radii also show a similar dependence (Fig. 28; Table 5, p. 36).

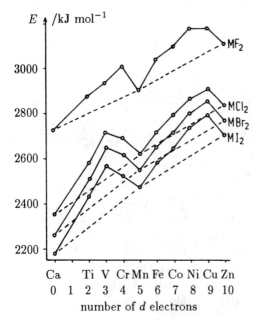

Fig. 27

Lattice energies of the dihalides of elements of the first transition metal period

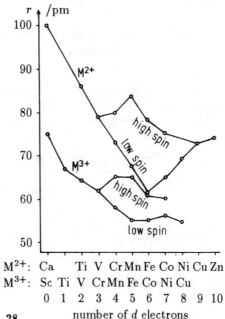

Fig. 28

Ionic radii of the elements of the first transition metal period in octahedral coordination

8.3 Coordination Polyhedra for Transition Metals

According to the preceding statements certain coordination polyhedra occur preferentially for compounds of transition metals, depending on the central atom, the oxidation state, and the kind of ligand. The general tendencies can be summarized as follows:

The series of 3d elements from scandium to iron as well as nickel preferably form octahedral complexes in the oxidation states I, II, III, and IV; octahedra and tetrahedra are known for cobalt, and tetrahedra for zinc and copper(I). Copper(II) (d^9) forms JAHN-TELLER distorted octahedra and tetrahedra. With higher oxidation states (= smaller ionic radii) and larger ligands the tendency to form tetrahedra increases. For vanadium(V), chromium(VI) and manganese(VII) almost only tetrahedral coordination is known (VF$_5$ is an exception). Nickel(II) low-spin complexes (d^8) can be either octahedral or square.

Among the heavier 4d and 5d elements, tetrahedral coordination only occurs for silver, cadmium, and mercury and when the oxidation states are very high as in ReO_4^- or OsO_4. Octahedra are very common, and higher coordination numbers, especially 7, 8, and 9, are not unusual, as for instance in ZrO_2 (c.n. 7), $Mo(CN)_8^{4-}$ or $LaCl_3$ (c.n. 9). A special situation arises for the electron configuration d^8, namely for Pd(II), Pt(II), Ag(III), and Au(III), which almost always have square coordination. Pd(0), Pt(0), Ag(I), Au(I), and Hg(II) (d^{10}) frequently show linear coordination (c.n. 2). In Table 12 the most important coordination polyhedra are summarized with corresponding examples.

Table 12 Most common coordination polyhedra for coordination numbers 2 to 6 for transition metal compounds

polyhedron	c.n.	electron config.	central atom	examples
linear arrangement	2	d^{10}	Cu(I), Ag(I), Au(I), Hg(II)	Cu_2O, $Ag(CN)_2^-$, $AuCN^*$, $AuCl_2^-$, $HgCl_2$, HgO^*
triangle	3	d^{10}	Cu(I), Ag(I), Au(I), Hg(II)	$Cu(CN)_3^{2-}$, $Ag_2Cl_5^{3-}$, $Au(PPh_3)_3^+$, HgI_3^-
square	4	d^8	Ni(II), Pd(II), Pt(II), Au(III)	$Ni(CN)_4^{2-}$, $PdCl_2^*$, PtH_4^{2-}, $Pt(NH_3)_2Cl_2$, $AuCl_4^-$
tetrahedron	4	d^0	Ti(IV), V(V), Cr(VI), Mo(VI), Mn(VII), Re(VII) Ru(VIII),Os(VIII)	$TiCl_4$, VO_4^{3-}, CrO_3^*, CrO_4^{2-}, MoO_4^{2-} Mn_2O_7, ReO_4^- RuO_4, OsO_4
		d^1	V(IV), Cr(V), Mn(VI), Ru(VII)	VCl_4, CrO_4^{3-}, MnO_4^{2-}, RuO_4^-
		d^5	Mn(II), Fe(III)	$MnBr_4^{2-}$, Fe_2Cl_6
		d^6	Fe(II)	$FeCl_4^{2-}$
		d^7	Co(II)	$CoCl_4^{2-}$
		d^8	Ni(II)	$NiCl_4^{2-}$
		d^9	Cu(II)	$CuCl_4^{2-}$ †
		d^{10}	Ni(0), Cu(I), Zn(II), Hg(II)	$Ni(CO)_4$, Cu_2O, $Zn(CN)_4^{2-}$, HgI_4^{2-}
square pyramid	5	d^0	Ti(IV), V(V), Nb(V), Mo(VI), W(VI),	$TiOCl_4^{2-}$, VOF_4^-, $NbSCl_4^-$, $MoNCl_4^-$, $WNCl_4^-$
		d^1	V(IV), Cr(V), Mo(V), W(V), Re(VI)	$VO(NCS)_4^{2-}$, $CrOCl_4^-$, $MoOCl_4^-$, $WSCl_4^-$, $ReOCl_4$
		d^2	Os(VI)	$OsNCl_4^-$
		d^4	Mn(III), Re(III)	$MnCl_5^{2-}$, Re_2Cl_8
		d^7	Co(II)	$Co(CN)_5^{3-}$
trigonal bipyramid	5	d^2	V(IV)	$VCl_3(NMe_3)_2$
		d^8	Fe(0)	$Fe(CO)_5$
octahedron	6		nearly all; rarely Pd(II), Pt(II), Au(III), Cu(I)	

* endless chain   † Jahn–Teller distorted

8.4 Isomerism

Two compounds are *isomers* when they have the same chemical composition but different molecular structures. Isomers have different physical and chemical properties.

Constitution isomers have molecules with different *constitutions*, i.e. the atoms linked with one another differ. For example:

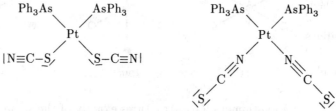

Transition metal complexes in particular show several kinds of constitution isomers, namely:

Bonding isomers, differing by the kind of ligand atom bonded to the central atom, for example:

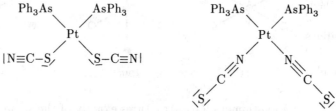

Further ligands that can be bonded by different atoms include OCN^- and NO_2^-. Cyanide ions always are linked with their C atoms in isolated complexes, but in polymer structures as in prussian blue they can be coordinated via both atoms ($Fe-C\equiv N-Fe$).

Coordination isomers occur when complex cations and complex anions are present and ligands are exchanged between anions and cations, for example:

$$[Cu(NH_3)_4][PtCl_4] \qquad [Pt(NH_3)_4][CuCl_4]$$
$$[Pt(NH_3)_4][PtCl_6] \qquad [Pt(NH_3)_4Cl_2][PtCl_4]$$

Further variations are:

Hydrate isomers, e.g. $[Cr(OH_2)_6]Cl_3$,
$[Cr(OH_2)_5Cl]Cl_2 \cdot H_2O$,
$[Cr(OH_2)_4Cl_2]Cl \cdot 2H_2O$

Ionization isomers, e.g. $[Pt(NH_3)_4Cl_2]Br_2$, $[Pt(NH_3)_4Br_2]Cl_2$

Stereo isomers have the same constitution, but a different spatial arrangement of their atoms; they differ in their *configuration.* Two cases have to be distinguished: geometrical isomers (diastereomers) and enantiomers.

Geometrical isomers occur as *cis-trans* isomers in compounds with double bonds like in N_2F_2 and especially when coordination polyhedra have different kinds of ligands. The most important types are square and octahedral complexes with two or more different ligands (Fig. 29). To designate them in more complicated cases, the polyhedron vertices are numbered alphabetically, for instance *abf*-triaqua-*cde*-tribromoplatinum(IV) for *mer*-$[PtBr_3(NH_3)_3]^+$. No geometrical isomers exist for tetrahedral complexes. With other coordination polyhedra the number of possible isomers increases with the number of different ligands (Table 13); however, usually only one or two of them are known.

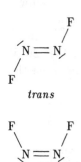

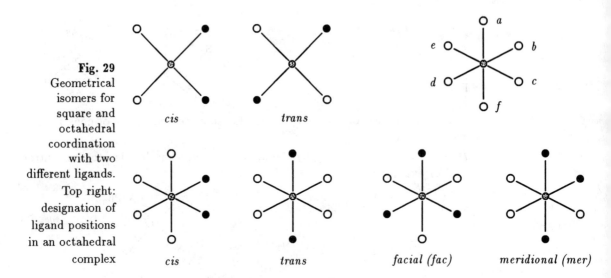

Fig. 29
Geometrical isomers for square and octahedral coordination with two different ligands. Top right: designation of ligand positions in an octahedral complex

cis   trans

facial (fac)   meridional (mer)

Enantiomers have structures exactly of the same kind and yet are not identical. Their structures correspond to mirror images. In their physical properties they only differ in respect to phenomena that are polar, i.e. that have some kind of a preferred directionality. This especially includes polarized light, the polarization plane of which experiences a rotation when it passes through a solution of the substance. For this reason enantiomers have also been called optical isomers. In their chemical properties enantiomers only differ when they react with a compound that is an enantiomer itself.

The requirement for the existence of enantiomers is a *chiral* structure. This condition is fulfilled when neither a mirror plane nor an inversion center nor an inversion axis is present as a symmetry element (a mirror plane and an inversion center can be regarded as special kinds of inversion axes; more details are given in chapter 18). The great majority of known chiral compounds are naturally ocurring organic substances, their molecules having one or more *asymmetrically substituted* C atoms. Chirality is present when a tetrahedrally coordinated atom has four different ligands.[*] Known inorganic enantiomers are mainly complex compounds, mostly with octahedral coordination [35]. In Table 13 ligand combinations are listed for which chiral molecules are possible. The best-known chiral complexes are chelate complexes, some examples being shown in Fig. 30. The configuration of trichelate complexes like $[Co(H_2N(CH_2)_2NH_2)_3]^{3+}$ can be designated by Δ or Λ: view the structure from the direction as in Fig. 30; if the chelate groups are oriented like the turns of a right-handed screw, then the symbol is Δ.

---

[*]In organic stereochemistry the term "center of chirality" is often used; usually it refers to an asymmetrically substituted C atom. This term should be avoided since it is a contradiction in itself: a chiral object by definition has no center (chirality is a symmetry property that explicitly excludes a center of symmetry).

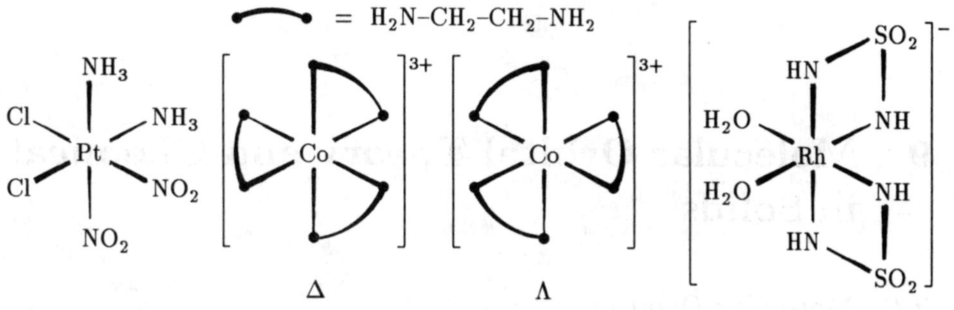

$$\text{———} = H_2N-CH_2-CH_2-NH_2$$

Fig. 30 Examples of some chiral complexes with octahedral coordination

Δ   Λ

Table 13 Number of possible geometrical isomers depending on the number of different ligands (designated by A, B, C, ...) for some coordination polyhedra (excluding chelate complexes). Of every pair of enantiomers only one representative was counted

polyhedron	ligands	total number	chiral number	polyhedron	ligands	total number	chiral number
tetrahedron	unrestricted	1	ABCD	octahedron	AB_5	1	0
square	AB_3	1	0		A_2B_4	2	0
	A_2B_2	2	0		A_3B_3	2	0
	ABC_2	2	0		ABC_4	2	0
	ABCD	3	0		AB_2C_3	3	0
trigonal	AB_4	2	0		$A_2B_2C_2$	5	1
bipyramid	A_2B_3	3	0		$ABCD_3$	4	1
	ABC_3	4	0		ABC_2D_2	6	2
	AB_2C_2	5	1		$ABCDE_2$	9	6
	$ABCD_2$	7	3		ABCDEF	15	15
	ABCDE	10	10				

8.5 Problems

8.1 State which of the following octahedral high-spin complexes should be JAHN–TELLER distorted.
TiF_6^{2-}, MoF_6, $[Cr(OH_2)_6]^{2+}$, $[Mn(OH_2)_6]^{2+}$, $[Mn(OH_2)_6]^{3+}$, $FeCl_6^{3-}$, $[Ni(NH_3)_6]^{2+}$, $[Cu(NH_3)_6]^{2+}$.

8.2 State which of the following tetrahedral complexes should be JAHN–TELLER distorted, and what kind of a distortion it should be.
$CrCl_4^-$, $MnBr_4^{2-}$, $FeCl_4^-$, $FeCl_4^{2-}$, $NiBr_4^{2-}$, $CuBr_4^{2-}$, $Cu(CN)_4^{3-}$, $Zn(NH_3)_4^{2+}$.

8.3 Decide whether the following complexes are tetrahedral or square.
$Co(CO)_4^-$, $Ni(PF_3)_4$, $PtCl_2(NH_3)_2$, $Pt(NH_3)_4^{2+}$, $Cu(OH)_4^{2-}$, Au_2Cl_6 (dimer via chloro bridges).

8.4 How many isomers do you expect for the following complexes?
(a) $PtCl_2(NH_3)_2$; (b) $ZnCl_2(NH_3)_2$; (c) $[OsCl_4F_2]^{2-}$; (d) $[CrCl_3(OH_2)_3]^{3-}$; (e) $Mo(CO)_5OR_2$.

9 Molecular Orbital Theory and Chemical Bonding in Solids

9.1 Molecular Orbitals

Molecular orbital (MO) theory currently offers the most precise description of the bonding within a molecule. The term *orbital* is a neologism reminiscent of the concept of an orbiting electron, but it also expresses the inadequacy of this concept for the precise characterization of the behavior of an electron. Mathematically an electron is treated as a standing wave by the formulation of a wave function ψ. For the hydrogen atom the wave functions for the ground state and all excited states are known exactly; they can be calculated as solutions of the SCHRÖDINGER equation. Hydrogen-like wave functions are assumed for other atoms, and their calculation is performed with approximation methods.

The wave function of an electron corresponds to the expression used to describe the amplitude of a vibrating chord as a function of the position x. The opposite direction of the motion of the chord on the two sides of a vibrational node is expressed by opposite signs of the wave function. Similarly, the wave function of an electron has opposite signs on the two sides of a nodal surface. The wave function is a function of the site x, y, z, referred to a coordinate system that has its origin in the center of the atomic nucleus.

Wave functions for the orbitals of molecules are calculated by linear combinations of *all* wave functions of *all* atoms involved. The total number of orbitals remains unaltered, i.e. the total number of contributing atomic orbitals must be equal to the number of molecular orbitals. Furthermore, certain conditions have to be obeyed in the calculation; these include linear independence of the molecular orbital functions and normalization. In the following we will designate wave functions of atoms by χ and wave functions of molecules by ψ. We obtain the wave functions of an H_2 molecule by linear combination of the $1s$ functions χ_1 and χ_2 of the two hydrogen atoms:

$$\psi_1 = \tfrac{1}{2}\sqrt{2}(\chi_1 + \chi_2) \qquad\qquad \psi_2 = \tfrac{1}{2}\sqrt{2}(\chi_1 - \chi_2)$$

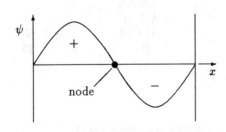

energy

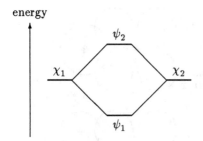

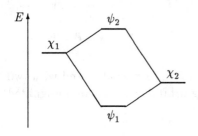

Compared to an H atom, electrons with the function ψ_1 are less energetic, and those with the function ψ_2 are more energetic. When the two available electrons "occupy" the molecular orbital ψ_1, this is energetically favorable; ψ_1 is the wave function of a *bonding* molecular orbital. ψ_2 belongs to an *antibonding* molecular orbital; its occupation by electrons requires the input of energy.

When calculating the wave functions for the bonds between two atoms of different elements, the functions of the atoms contribute with different coefficients c_1 and c_2:

$$\psi_1 = c_1\chi_1 + c_2\chi_2 \qquad (8)$$
$$\psi_2 = c_2\chi_1 - c_1\chi_2 \qquad (9)$$

The probability of finding an electron at a site x, y, z is given by ψ^2. Integrated over all space, the probability must be equal to 1:

$$1 = \int \psi_1^2\, dV = \int |c_1\chi_1 + c_2\chi_2|^2\, dV = c_1^2 + c_2^2 + 2c_1c_2S_{12} \qquad (10)$$

S_{12} is the *overlap integral* between χ_1 and χ_2. The term $2c_1c_2S_{12}$ is the *overlap population*; it expresses the electronic interaction between the atoms. The contributions c_1^2 and c_2^2 can be assigned to the atoms 1 and 2, respectively.

Equation (10) is fulfilled when $c_1^2 \approx 1$ and $c_2^2 \approx 0$; in this case the electron is localized essentially at atom 1 and the overlap population is approximately zero. This is the situation of a minor electronic interaction, either because the corresponding orbitals are too far apart or because they differ considerably in energy. Such an electron does not contribute to bonding.

For ψ_1 the overlap population $2c_1c_2S_{12}$ is positive, and the electron is bonding; for ψ_2 it is negative, and the electron is antibonding. The sum of the values $2c_1c_2S_{12}$ of all occupied orbitals of the molecule, the MULLIKEN overlap population, is a measure for the bond strength or bond order (b.o.); b.o. = $\frac{1}{2}$[(number of bonding electrons) − (number of antibonding electrons)].[*]

Orbitals other than s orbitals can also be combined to bonding, antibonding or nonbonding molecular orbitals. Nonbonding are those orbitals for which bonding and antibonding components cancel each other. Some possibilities are shown in Fig. 31. Note the signs of the wave functions. A bonding molecular orbital having no nodal surface is a σ orbital; if it has one nodal plane parallel

---

[*]Despite the given formula, the calculation of the bond order is not always clear, because orbitals having only a minor bonding or minor antibonding effect are not always considered. Nevertheless, the bond order is a simple and useful concept. Chemists very successfully use many concepts in a more intuitive manner, although they generally tend not to define their concepts clearly, and to ignore definitions.

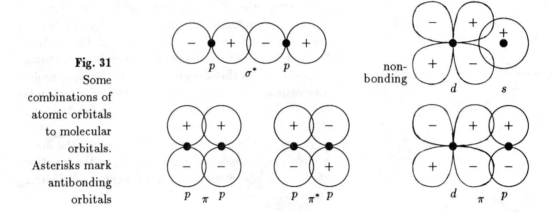

Fig. 31
Some
combinations of
atomic orbitals
to molecular
orbitals.
Asterisks mark
antibonding
orbitals

to the connecting line between the atomic centers it is a π orbital, and with two such nodal planes it is a δ orbital. Antibonding orbitals usually are designated by an asterisk *.

9.2 Hybridization

In order to calculate the orbitals for a methane molecule, the four $1s$ functions of the four hydrogen atoms and the functions $2s$, $2p_x$, $2p_y$ and $2p_z$ of the carbon atom are combined to give eight wave functions, four of which are bonding and four of which are antibonding. The four bonding wave functions are:

$$\psi_1 = \frac{c_1}{2}(s + p_x + p_y + p_z) + c_2\chi_{H1} + c_3(\chi_{H2} + \chi_{H3} + \chi_{H4})$$

$$\psi_2 = \frac{c_1}{2}(s + p_x - p_y - p_z) + c_2\chi_{H2} + c_3(\chi_{H3} + \chi_{H4} + \chi_{H1})$$

$$\psi_3 = \frac{c_1}{2}(s - p_x + p_y - p_z) + c_2\chi_{H3} + c_3(\chi_{H4} + \chi_{H1} + \chi_{H2})$$

$$\psi_4 = \frac{c_1}{2}(s - p_x - p_y + p_z) + c_2\chi_{H4} + c_3(\chi_{H1} + \chi_{H2} + \chi_{H3})$$

ψ_1, ψ_2,... are wave functions of the CH_4 molecule, s, p_x, p_y and p_z designate the wave functions of the C atom, and χ_{H1}, χ_{H2},... correspond to the H atoms. Among the coefficients c_1, c_2 and c_3, one is negligible: $c_3 \approx 0$.

Formulated as in the preceding paragraph, the functions are not especially illustrative. They do not correspond to the idea a chemist associates with the formation of a bond between two atoms: in his or her imagination the atoms approach each other and their atomic orbitals merge into a bonding molecular orbital. To match this kind of mental picture it is expedient to start from atomic orbitals whose spatial orientations correspond to the orientations of the bonds of the molecule that is formed. Such orbitals can be obtained by "hybridization" of atomic orbitals. Instead of calculating the molecular orbitals of the methane molecule in one step according to the equations mentioned above, one proceeds in two steps. First, only the wave functions of the C atom are combined to sp^3 hybrid orbitals:

$$\chi_1 = \frac{1}{2}(s + p_x + p_y + p_z)$$

$$\chi_2 = \frac{1}{2}(s + p_x - p_y - p_z)$$

$$\chi_3 = \frac{1}{2}(s - p_x + p_y - p_z)$$

$$\chi_4 = \frac{1}{2}(s - p_x - p_y + p_z)$$

The functions χ_1 to χ_4 correspond to orbitals having preferential directionalities oriented towards the vertices of an circumscribed tetrahedron. Their combinations with the wave functions of four hydrogen atoms placed in these vertices yield the following functions, the insignificant coefficient c_3 being neglected:

$$\psi_1 = c_1\chi_1 + c_2\chi_{H1}$$
$$\psi_2 = c_1\chi_2 + c_2\chi_{H2}$$
$$\text{etc.}$$

ψ_1 corresponds to a bonding orbital that essentially involves the interaction of the C atom with the first H atom; its charge density ψ_1^2 is concentrated in the region between these two atoms. This matches the idea of a localized C–H bond. To be more exact, every bond is a "multi-center bond" with contributions of the wave functions of all atoms; however, due to the charge concentration in the region between two atoms and because of the inferior contributions χ_{H2}, χ_{H3}, and χ_{H4}, the bond can be taken in good approximation to be a "two-electron-two-center bond" ($2e2c$ bond) between the atoms C and H1. The hybridization is not necessary for the calculation; it is, however, a helpful mathematical trick for adapting the wave functions to a chemist's mental picture.

For molecules with different structures different hybridization functions are appropriate. An infinity of hybridization functions can be formulated by linear combinations of s and p orbitals:

$$\chi_i = \alpha_i s + \beta_i p_x + \gamma_i p_y + \delta_i p_z$$

The coefficients must be normalized, i.e. $\alpha_i^2 + \beta_i^2 + \gamma_i^2 + \delta_i^2 = 1$. Their values determine the preferential directions of the hybrid orbitals. For example, the functions

$$\chi_1 = 0.833s + 0.32(p_x + p_y + p_z)$$
$$\chi_2 = 0.32s + 0.547(p_x - p_y - p_z)$$
$$\chi_3 = 0.32s + 0.547(-p_x + p_y - p_z)$$
$$\chi_4 = 0.32s + 0.547(-p_x - p_y + p_z)$$

define an orbital (χ_1) having contributions of 69 % ($=0.833^2 \times 100$ %) s and 31 % p and three orbitals (χ_2, χ_3, χ_4), each with contributions of 10 % s and 90 % p. They are adequate to calculate the wave functions for a molecule $|AX_3$ that has a lone electron pair (χ_1) with a larger s contribution and bonds with larger

p orbital contributions as compared to sp^3 hybridization. The corresponding bond angles are between $90°$ and $109.5°$, namely $96.5°$.

To derive the values of the coefficients α_i, β_i, γ_i, and δ_i so that the bond energy is maximized and the correct molecular geometry results, the mutual interactions between the electrons have to be considered. This requires a great deal of computational expenditure. However, in a qualitative manner the interactions can be estimated rather well: that is exactly what the valence shell electron-pair repulsion theory accomplishes.

9.3   Band Theory. The Linear Chain of Hydrogen Atoms

In a solid that cannot be interpreted on the base of localized covalent bonds or of ions, the assessment of the bonding requires the consideration of the complete set of molecular orbitals of *all* involved atoms. This is the subject of *band theory*, which offers the most comprehensive conception of chemical bonding. Ionic bonding and localized covalent bonds result as special cases. The ideas presented in this chapter are based on the intelligible exposition by R. HOFFMANN [37], the reading of which is recommended for a deeper insight into the subject. To begin with, we regard a linear chain of $N + 1$ evenly spaced hydrogen atoms. By the linear combination of their $1s$ functions we obtain $N + 1$ wave functions $\psi_{k'}$; $k' = 0, \ldots, N$. The wave functions have some similarity to the standing waves of a vibrating chord or, better, with the vibrations of a chain of $N + 1$ spheres that are connected by springs (Fig. 32). The chain can adopt different vibrational modes that differ in the number of vibrational nodes; we number the modes by sequential numbers k' corresponding to the number of nodes. k' cannot be larger than N, as the chain cannot adopt more nodes than spheres. We number the $N + 1$ spheres from $n = 0$ to $n = N$. Every sphere vibrates with a certain amplitude:

$$A_n = A_0 \cos 2\pi \frac{k'n}{2N}$$

Every one of the standing waves has a wavelength $\lambda_{k'}$:

$$\lambda_{k'} = \frac{2Na}{k'}$$

a is the distance between two spheres. Instead of numbering the vibrational modes with sequential numbers k', it is more convenient to use wave numbers k:

$$k = \frac{2\pi}{\lambda_{k'}} = \frac{\pi k'}{Na}$$

This way one becomes independent of the number N, as the limits for k become 0 and π/a. Contrary to the numbers k' the values for k are not integral numbers.

The k-th wave function of the electrons in a chain of hydrogen atoms results in a similar way. From every atom we obtain a contribution $\chi_n \cos nka$, i.e. the $1s$ function χ_n of the n-th atom of the chain takes the place of A_0. All atoms have the same function χ, referred to the local coordinate system of the atom,

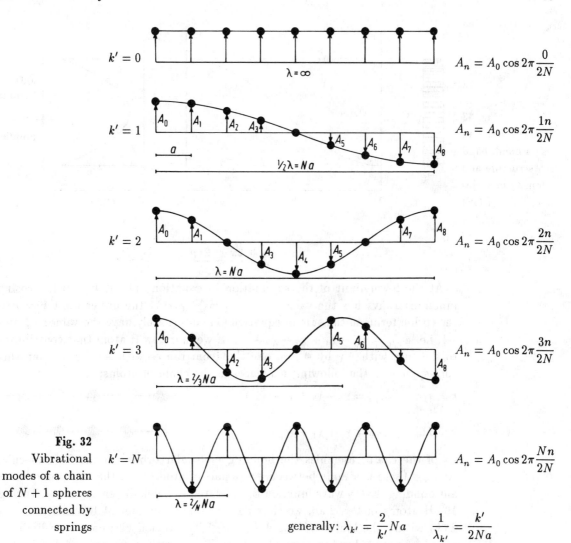

$$A_n = A_0 \cos 2\pi \frac{0}{2N}$$

$$A_n = A_0 \cos 2\pi \frac{1n}{2N}$$

$$A_n = A_0 \cos 2\pi \frac{2n}{2N}$$

$$A_n = A_0 \cos 2\pi \frac{3n}{2N}$$

$$A_n = A_0 \cos 2\pi \frac{Nn}{2N}$$

Fig. 32
Vibrational modes of a chain of $N+1$ spheres connected by springs

generally: $\lambda_{k'} = \dfrac{2}{k'} Na \qquad \dfrac{1}{\lambda_{k'}} = \dfrac{k'}{2Na}$

and the index n designates the position of the atom in the chain. The k-th wave function is composed of contributions of all atoms:

$$\psi_k = \sum_{n=0}^{N} \chi_n \cos nka \qquad (11)$$

A wave function composed this way from the contributions of single atoms is called a BLOCH function (in texts on quantum chemistry you will find this function being formulated with exponential functions $\exp(inka)$ instead of the cosine functions, since this facilitates the mathematical treatment).

The number k is more than just a simple number to designate a wave function. According to the DE BROGLIE equation $p = h/\lambda$ every electron can be assigned a momentum p (h = PLANCK constant). k and the momentum are related:

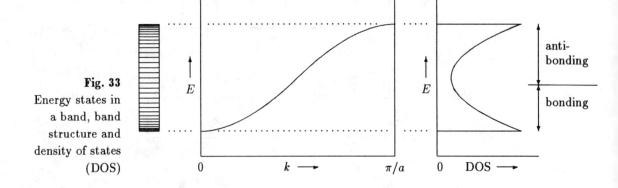

Fig. 33
Energy states in
a band, band
structure and
density of states
(DOS)

$$k = \frac{2\pi}{\lambda} = \frac{2\pi p}{h} \tag{12}$$

At the lower limit of the summation in equation (11) at $k = 0$ the cosine function always has the value 1, i.e. $\psi_0 = \sum \chi_n$. At the upper limit $k = \pi/a$ the cosine terms in the sum of equation (11) alternately have the values $+1$ and -1, i.e. $\psi_{\pi/a} = \chi_0 - \chi_1 + \chi_2 - \chi_3 + \ldots$. If we mark an H atom that contributes to the sum with $+\chi$ by ● and one that contributes with $-\chi$ by ○, then this corresponds to the following sequences in the chain of atoms:

$k = \pi/a:$ $\psi_{\pi/a} = \chi_0 - \chi_1 + \chi_2 - \chi_3 + \cdots$ ●—○—●—○—●

$k = 0:$ $\psi_0 = \chi_0 + \chi_1 + \chi_2 + \chi_3 + \cdots$ ●—●—●—●—●

ψ_0 of the chain resembles the bonding molecular orbital of the H_2 molecule. At $\psi_{\pi/a}$ there is a node between every pair of atoms, and the wave function is antibonding. Every wave function ψ_k is related to a definite energy state. Taking 10^6 H atoms in the chain we thus have the huge number of 10^6 energy states $E(k)$ within the limits $E(0)$ and $E(\pi/a)$.[*] The region between these limits is called an *energy band* or a *band* for short. The energy states are not distributed evenly in the band. Fig. 33 shows on the left a scheme of the band in which every line represents one energy state; only 38 instead of 10^6 lines were drawn. In the center the *band structure* is plotted, i.e. the energy as a function of k; the curve is not really continuous as it appears but consists of numerous tightly crowded dots, one for each energy state. The curve flattens out at the ends, showing a denser sequence of the energy levels at the band limits. The *density of states* (DOS) is shown at the right side; DOS·dE = number of energy states between E and $E+dE$. The energy levels in the lower part of the band belong to bonding states, and in the upper part to antibonding states.

The *band width* or *band dispersion* is the energy difference between the highest and the lowest energy level in the band. The band width becomes larger when the interaction among the atoms increases, i.e. when the atomic orbitals overlap

---

[*]10^6 atoms with interatomic distances of 100 pm can be accomodated in a chain of 0.1 mm length

to a greater extent. A smaller interatomic distance a causes a larger band width. For the chain of hydrogen atoms a band width of 4.4 eV is calculated when adjacent atoms are separated by 200 pm, and 39 eV results when they move up to 100 pm.

According to the PAULI principle two electrons can adopt the same wave function, so that the N electrons of the N H atoms take the energy states in the lower half of the band, and the band is "half occupied". The highest occupied energy level ($=$ HOMO $=$ highest occupied molecular orbital) is the *Fermi limit*. Whenever the FERMI limit is inside a band, metallic electric conduction is observed. Only a very minor energy supply is needed to promote an electron from an occupied state under the FERMI limit to an unoccupied state above it; the easy switchover from one state to another is equivalent to a high electron mobility. Because of excitation by thermal energy a certain fraction of the electrons is always found above the FERMI limit.

The curve for the energy dependence as a function of k in Fig. 33 has a positive slope. This is not always so. When p orbitals are joined head-on to a chain, the situation is exactly the opposite. The wave function $\psi_0 = \sum \chi_n$ is then antibonding, whereas $\psi_{\pi/a}$ is bonding (Fig. 34).

Different bands can overlap each other, i.e. the lower limit of one band can have a lower energy level than the upper limit of another band. This applies especially for wide bands.

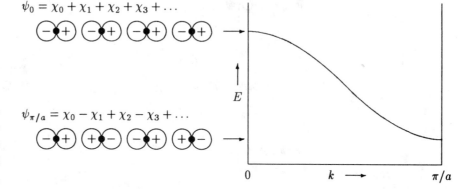

Fig. 34
Band structure
for a chain of p
orbitals oriented
head-on

$$\psi_0 = \chi_0 + \chi_1 + \chi_2 + \chi_3 + \cdots$$

$$\psi_{\pi/a} = \chi_0 - \chi_1 + \chi_2 - \chi_3 + \cdots$$

9.4   The Peierls Distortion

The model of the chain of hydrogen atoms with a completely delocalized (metallic) type of bonding is outlined in the preceding section. Intuitively, a chemist will find this model rather unreal, as he or she expects the atoms to combine pairwise to H_2 molecules. In other words, the chain of equidistant H atoms is expected to be unstable, so it undergoes a distortion in such a way that the atoms approach each other in pairs. This process is called PEIERLS distortion (or strong electron–phonon coupling) in solid state physics:

$$\cdots\text{H}\cdots\cdots\text{H}\cdots\cdots\text{H}\cdots\cdots\text{H}\cdots\cdots\text{H}\cdots\cdots\text{H}\cdots$$

$$\downarrow$$

$$\text{H}—\text{H}\qquad\text{H}—\text{H}\qquad\text{H}—\text{H}$$

The very useful chemist's intuition, however, is of no help when the question arises of how hydrogen will behave at a pressure of 10^8 bar. Presumably it will be metallic then.

Let us consider once more the chain of hydrogen atoms, but this time we put it together starting from H_2 molecules. In the beginning the chain then consists of H atoms, and electron pairs occur between every other pair of atoms. Nevertheless, let us still assume equidistant H atoms. The orbitals of the H_2 molecules interact with one another to give a band. As the repeating unit, i.e. the lattice constant in the chain, is now doubled to $2a$, the k values only run from $k = 0$ to $k = \pi/(2a)$. Instead, we have two branches in the curve for the band energy (Fig. 35). One branch begins at $k = 0$ and has a positive slope; it starts from the bonding molecular orbitals of the H_2, all having the same sign for their wave functions. The second branch starts at $k = 0$ with the higher energy of the antibonding H_2 orbital and has a negative slope. Both branches meet at $k = \pi/(2a)$, with two equivalent, i.e. degenerate, wave functions.

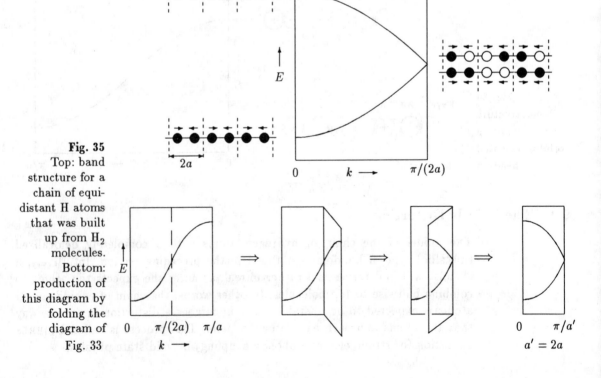

Fig. 35
Top: band structure for a chain of equidistant H atoms that was built up from H_2 molecules. Bottom: production of this diagram by folding the diagram of Fig. 33

As a result, the same band structure must result for the H atom chain, irrespective whether it is based on the wave functions of N H atoms or of $N/2$ H_2 molecules. As a matter of fact, the curve of Fig. 33 coincides with the curve of Fig. 35. The apparent difference has to do with the doubling of the lattice constant from a to $a' = 2a$. As we see from equation (11), the same wave functions ψ_k result for $k = 0$ and for $k = 2\pi/a$, the same ones for $k = \pi/a$ and for $k = 3\pi/a$, etc. Whereas the curve in Fig. 33 runs steadily upwards from $k = 0$ to $k = \pi/a$, in Fig. 35 it only runs until $k = \pi/(2a) = \pi/a'$, then it continues upwards from right to left. We can obtain the one plot from the other by folding the diagram, as is shown in the lower part of Fig. 35. The folding can be continued: triplication of the unit cell requires two folds, etc.

Up to now we have assumed evenly spaced H atoms. If we now allow the H atoms to approach each other pairwise, a change in the band structure takes place. The corresponding movements of the atoms are marked by arrows in Fig. 35. At $k = 0$ this has no consequences; at the lower (or upper) end of the band an energy gain (or loss) occurs for the atoms that approach each other; it is compensated by the energy loss (or gain) of the atoms moving apart. However, in the central part of the band, where the H atom chain has its FERMI limit, substantial changes take place. One of the degenerate states is stabilized while the other one is destabilized. The upper branch of the curve shifts upwards, and the lower one downwards. As a result a gap opens up, and the band splits (Fig. 36). For the half-filled band the net result is an energy gain. Therefore, it is energetically more favorable when short and long distances between the H atoms alternate in the chain. The chain no longer is an electrical conductor, as an electron must overcome the energy gap in order to pass from one energy state to another.

The one-dimensional chain of hydrogen atoms is merely a model. However, compounds do exist for which the same kind of considerations apply and have

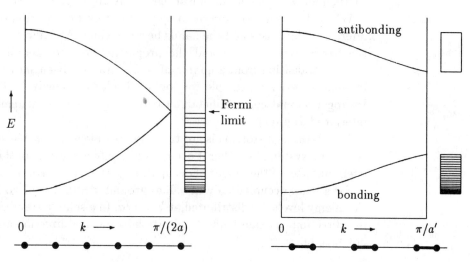

Fig. 36 Band structure for a chain of H atoms. Left, with equidistant atoms; right, after PEIERLS distortion to H_2 molecules. The lines in the rectangles symbolize energy states occupied by electrons

been confirmed experimentally. These include polyene chains such as polyacetylene. The p orbitals of the C atoms take the place of the $1s$ functions of the H
atoms; they form one bonding and one antibonding π band. Due to the PEIERLS
distortion the polyacetylene chain is only stable with alternate short and long
C–C bonds, that is, in the sense of the valence bond formula with alternate
single and double bonds:

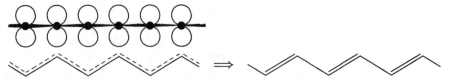

Polyacetylene is not an electrical conductor. If it is doped with an impurity that
either introduces electrons into the upper band or that removes electrons from
the lower band, it becomes a good conductor.

The PEIERLS distortion is a substantial factor influencing which structure a
solid adopts. The driving force is the tendency to maximize bonding, i.e. the
same tendency that forces H atoms or other radicals to bond with each other.
In a solid, that amounts to shifting the density of states at the FERMI level,
in that bonding states are shifted towards lower and antibonding states towards higher energy values. By opening up an energy gap the bands become
narrower; within a band the energy levels become more crowded. The extreme
case is a band that has shrunk to a single energy value, i.e. all levels have the
same energy. This happens, for example, when the chain of hydrogen atoms
consists of widely separated H_2 molecules; then we have separate, independent
H_2 molecules whose energy levels all have the same value; the bonds are localized in the molecules. Generally, the band width is a measure for the degree of
localization of the bonds: a narrow band represents a high degree of localization, and with increasing band width the bonds become more delocalized. Since
narrow bands can hardly overlap and are usually separated by intervening gaps,
compounds with essentially localized bonds are electrical insulators.

When the atoms are forced to move closer by the exertion of pressure, their
interaction increases and the bands become wider. At sufficiently high pressures
the bands overlap again and the properties become metallic. The pressure-
induced transition from a nonmetal to a metal has been shown experimentally
in some cases, for example for iodine. Under extremely high pressures even
hydrogen should become metallic (metallic hydrogen is assumed to exist in the
interior of Jupiter).

The PEIERLS distortion is not the only possible way to achieve the most stable
state for a system. Whether it occurs is not only a question of the band structure
itself, but also of the degree of occupation of the bands. For an unoccupied band
or for a band occupied only at values around $k = 0$, it is of no importance how
the energy levels are distributed at $k = \pi/a$. In a solid, a stabilizing distortion in
one direction can cause a destabilization in another direction and may therefore
not take place.

9.5   Crystal Orbital Overlap Population (COOP)

At the end of section 9.1 the MULLIKEN overlap population is mentioned as a quantity related to the bond order. A corresponding quantity for solids was introduced by R. HOFFMANN: the *crystal orbital overlap population* (COOP). It is a function that specifies the bond strength in a crystal, all states being taken into account by the MULLIKEN overlap populations $2c_i c_j S_{ij}$. Its calculation requires a powerful computer; however, it can be estimated in a qualitative manner by considering the interactions between neighboring atomic orbitals, such as shown in Fig. 37. At $k = 0$ all interatomic interactions are bonding. At $k = \pi/a$ they are antibonding for directly adjacent atoms, but they are bonding between every other atom, albeit with reduced contributions due to the longer distance. At $k = \pi/(2a)$ the contributions between every other atom are antibonding, and those of adjacent atoms cancel each other. By also taking into account the densities of states one obtains the COOP diagram. In it net bonding overlap populations are plotted to the right and antibonding ones to the left. By marking the FERMI level it can be discerned to what extent bonding interactions predominate over antibonding interactions: they correspond to the areas enclosed by the curve below the FERMI level to the right and left side, respectively.

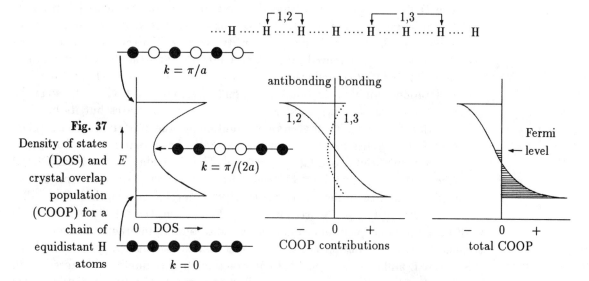

Fig. 37 Density of states (DOS) and crystal overlap population (COOP) for a chain of equidistant H atoms

Even in more complicated cases it is possible to obtain a qualitative idea. We choose the example studied by R. HOFFMANN of planar PtX_4^{2-} units that form a chain with Pt–Pt contacts. This kind of a structure is found for $K_2Pt(CN)_4$ and its partially oxidized derivates like $K_2Pt(CN)_4Cl_{0.3}$[*]:

---

[*]In the oxidized species the ligands have a staggered arrangement along the chain, but this is of no significant importance for our considerations

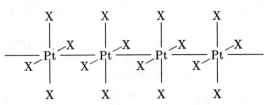

We will consider only the Pt–Pt interactions within the chain in the following. Fig. 38 shows the orientations of the relevant atomic orbitals at $k = 0$ and $k = \pi/a$. Aside from d orbitals one p orbital is also taken into account. At the lower left is plotted the sequence of the energy states of the molecular orbitals of the monomer, square complex (cf. Fig. 25, p. 65). The sketch to its right indicates how the energy levels fan out into bands when the PtX_4^{2-} ions are joined to a chain. The bands become wider the more intensely the orbitals interact with one another. With the aid of the orbital representations in the upper part of the figure the differences can be estimated: the orbitals d_{z^2} and p_z are oriented towards each other, and produce the widest bands; the interaction of the orbitals d_{xz} and d_{yz} is lower, and for d_{xy} and $d_{x^2-y^2}$ it is rather small (the band width for $d_{x^2-y^2}$ is slightly larger than for d_{xy} because of the inflation of $d_{x^2-y^2}$ due to its interaction with the ligands). The central plot shows the band structure, and the one on the right the density of states.

The DOS diagram results from the superposition of the densities of states of the different bands (Fig. 39). The d_{xy} band is narrow, its energy levels are crowded, and therefore it has a high density of states. For the wide d_{z^2} band the energy levels are distributed over a larger interval, and the density of states is smaller. The COOP contribution of every band can be estimated. This requires consideration mainly of the bonding action (overlap population), but also of the density of states. The d_{z^2} band has a lower density of states but its bonding interaction is strong, so that its contribution to the COOP is considerable. The opposite applies for the d_{xy} band. Generally, broad bands contribute more to the crystal orbital overlap population. The diagram for the total COOP at the bottom right of Fig. 39 results from the superposition of the COOP contributions of the different bands; the FERMI level is also marked. Since all d orbitals except $d_{x^2-y^2}$ are occupied in the PtX_4^{2-} ion, the corresponding bands also are fully occupied, and bonding and antibonding interactions compensate each other. By oxidation antibonding electrons are removed, the FERMI limit is lowered, and the bonding Pt–Pt interactions predominate. This agrees with observations: in $K_2Pt(CN)_4$ and similar compounds the Pt–Pt distances are about 330 pm; in the oxidized derivates $K_2Pt(CN)_4X_x$ they are shorter (270 to 300 pm, depending on the value of x; $X = Cl^-$ etc.).

9.6   Bonds in Two and Three Dimensions

In principle, the calculation of bonding in two or three dimensions follows the same scheme as outlined for the one-dimensional chain. Instead of one lattice

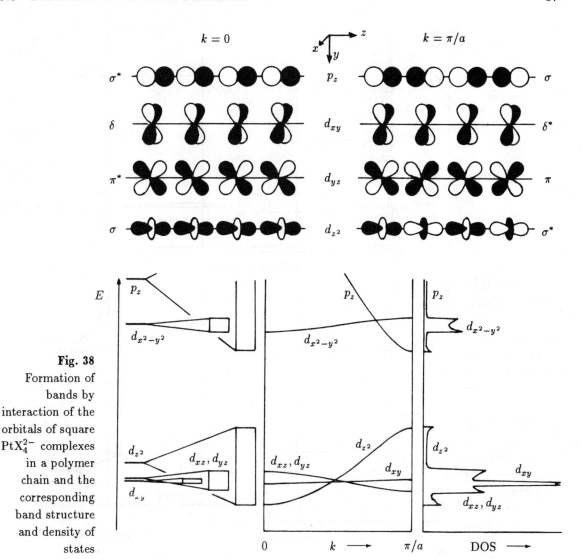

Fig. 38
Formation of
bands by
interaction of the
orbitals of square
PtX$_4^{2-}$ complexes
in a polymer
chain and the
corresponding
band structure
and density of
states

constant a, two or three lattice constants a, b and c have to be considered,
and instead of one sequential number k, two or three numbers k_x, k_y and k_z
are needed. The triplet of numbers $\mathbf{k} = (k_x, k_y, k_z)$ is called *wave vector*. This
term expresses the relation with the momentum of the electron. The momen-
tum has vectorial character, its direction coinicides with the direction of $\mathbf{k}$; the
magnitudes of both are related by the DE BROGLIE relation (equation (12)). In
the directions $\mathbf{a}$, $\mathbf{b}$ and $\mathbf{c}$ the components of $\mathbf{k}$ run from 0 to π/a, π/b and π/c,
respectively. As the direction of motion and the momentum of an electron can
be reversed, we also allow for negative values of k_x, k_y and k_z, with values that
run from 0 to $-\pi/a$ etc. However, for the calculation of the energy states the
positive values are sufficient, since according to equation (11) the energy of a
wave function is $E(k) = E(-k)$.

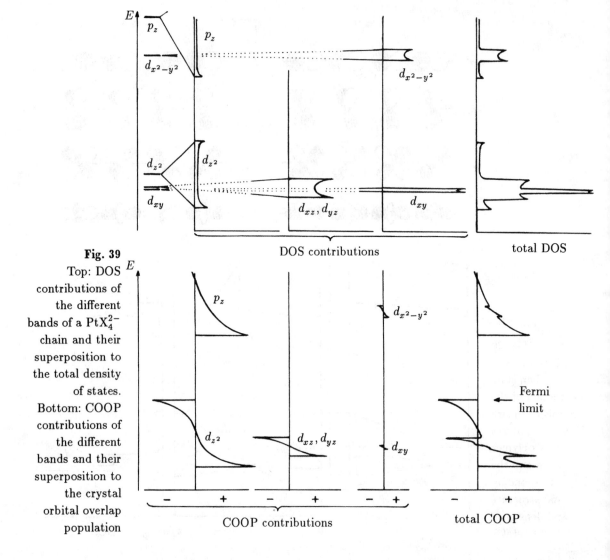

Fig. 39
Top: DOS contributions of the different bands of a PtX_4^{2-} chain and their superposition to the total density of states. Bottom: COOP contributions of the different bands and their superposition to the crystal orbital overlap population

The magnitude of **k** corresponds to a wave number $2\pi/\lambda$ and therefore is measured with a unit of reciprocal length. For this reason **k** is said to be a vector in a "reciprocal space" or "k space".* This is a "space" in a mathematical sense, i.e. it is concerened with vectors in a coordinate system, the axes of which serve to plot k_x, k_y and k_z. The directions of the axes run perpendicular to the delimiting faces of the unit cell of the crystal.

The region within which **k** is considered $(-\pi/a \leq k_x \leq \pi/a$ etc.) is the *first Brillouin zone*. In the coordinate system of k space it is a polyhedron. The faces of the first BRILLOUIN zone are oriented perependicular to the directions from one atom to the equivalent atoms in the adjacent unit cells. The distance of a

_____

*Compared to the reciprocal space commonly used in crystallography the k space is expanded by a factor 2π, otherwise the construction for both is the same.

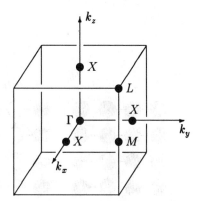

Fig. 40
First Brillouin zone for a cubic-primitive
crystal lattice. The points X are located
at $k = \pi/a$ in each case

face from the origin of the k coordinate system is π/s, s being the distance between the atoms. The first BRILLOUIN zone for a cubic-primitive crystal lattice
is shown in Fig. 40; the symbols commonly given to certain points of the BRIL-
LOUIN zone are labeled. The BRILLOUIN zone consists of a very large number
of small cells, one for every electronic state.

The pictures in Fig. 41 give an impression of how different s orbitals interact
with each other in a square lattice. Depending on the k values, i.e. for different
points in the BRILLOUIN zone, different kinds of interactions result. Between
adjacent atoms there are only bonding interactions at Γ, and only antibonding
interactions at M; the wave function corresponding to Γ therefore is the most
favorable one energetically, and the one corresponding to M the least favorable.
At X every atom has two bonding and two antibonding interactions with adjacent atoms, and its energy level is intermediate between that of Γ and M. It
is hardly possible to visualize the energy levels for all of the BRILLOUIN zone,
but one can plot diagrams that show how the energy values run along certain
directions within the zone. This has been done in the lower part of Fig. 41 for
three directions ($\Gamma \to X$, $X \to M$ and $\Gamma \to M$).

p_z orbitals that are oriented perpendicular to the square lattice interact in
the same way as the s orbitals, but the π-type interactions are inferior and
correspondingly the band width is smaller. For p_x and p_y orbitals the situation is
somewhat more complicated, because σ and π interactions have to be considered
between adjacent atoms (Fig. 41). For instance, at Γ the p_x orbitals are σ-
antibonding, but π-bonding. At X p_x and p_y differ most, one being σ and
π-bonding, and the other σ and π-antibonding.

In a three-dimensional, cubic-primitive lattice (α-Po structure, Fig. 4, p. 8)
the situation is similar. By stacking square nets and considering how the orbitals
interact at different points of the BRILLOUIN zone, a qualitative picture of the
band structure can be obtained.

9.7 Bonding in Metals

The density of states for the elements of a long period of the periodic table can
be sketched roughly as in Fig. 42. Due to the three-dimensional structure the

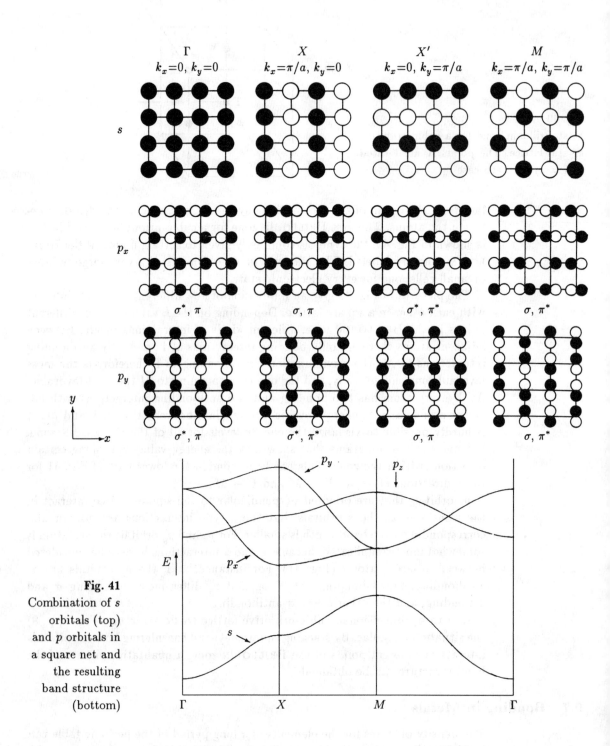

Fig. 41 Combination of s orbitals (top) and p orbitals in a square net and the resulting band structure (bottom)

more accurate consideration will no longer yield the simple DOS curves with two peaks as for a linear chain, but yields instead more or less complicated curves with numerous peaks. We will not go into the details here; in Fig. 42 merely a rectangle represents the DOS curve of each band. In each case the lower part of a band is bonding, and the upper part is antibonding. Correspondingly, the COOP diagram shows a contribution to the left and to the right side for every band; in the case of the p band as a whole there are more antibonding than bonding contributions so that its left side predominates. In the series potassium, calcium, scandium, ... we add a valence electron from element to element, and the FERMI limit climbs; the FERMI limit is marked to the right side of the figure for some valence electron counts. As can be seen, at first bonding states are occupied and therefore the bond strength increases for the metals from potassium to chromium. For the seventh to tenth valence electrons only antibonding states are available, and so the bond strengths decrease from chromium to nickel. The next electrons (Cu, Zn) are weakly bonding. With more than 14 valence electrons the total overlap population for a metallic structure becomes negative; structures with lower coordination numbers become favored.

The outlined sketch is rather rough, but it correctly shows the tendencies, as can be exemplified by the melting points of the metals (values in °C):

K	Ca	Sc	Ti	V	Cr	Mn	Fe	Co	Ni	Cu	Zn
63	839	1539	1667	1915	1900	1244	1535	1495	1455	1083	420

In reality there are subtle deviations from this simple picture. The energy levels shift somewhat from element to element, and different structure types have different band structures that become more or less favorable depending on the valence electron concentration. Furthermore, in the COOP diagram of Fig. 42 the s–p, s–d and p–d interactions were not taken into account, although they cannot be neglected. A more exact calculation shows that only antibonding contributions are to be expected from the eleventh valence electron onwards [37].

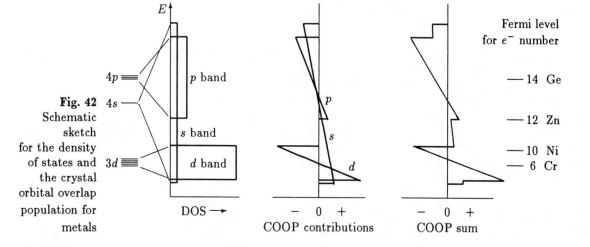

Fig. 42 Schematic sketch for the density of states and the crystal orbital overlap population for metals

9.8 Problems

9.1 What changes should occur in the band structure and the DOS diagrams (Fig. 33) when the chain of H atoms is compressed?

9.2 What would the band structure of a chain of p orbitals oriented head-on (Fig. 34) look like after a PEIERLS distortion?

9.3 What changes should occur in the band structure of the square net (Fig. 41) when it is compressed in the x direction?

10   The Element Structures of the Nonmetals

The $8 - N$ rule is presented in chapter 7 (p. 49). It states that an atom X of an element of the N-th main group of the periodic table will participate in $8 - N$ covalent bonds ($N = 4$ to 7):

$$b(XX) = 8 - N$$

In addition, as a rule, the *principle of maximal connectivity* holds for elements of the third and higher periods: the $8-N$ bonds usually are bonds to $8-N$ *different* atoms, and multiple bonds are avoided. For carbon, however, being an element of the second period, the less connected graphite is more stable than diamond at normal conditions. At higher pressures the importance of the principle of maximal connectivity increases; then, diamond becomes more stable. Even for nitrogen and oxygen the occurence of polymer modifications is to be expected at extremely high pressures.*

10.1   Halogens

Fluorine, chlorine, bromine and iodine consist of molecules X_2, even when in the solid state. In α-F_2 the F_2 molecules are packed in hexagonal layers; the molecules are oriented perpendicular to the layer, and the layers are stacked in the same way as in cubic closest-packing. Above 45.6 K up to the melting point (53.5 K) the modification β-F_2 is stable in which the molecules rotate about their centers of gravity.

The molecules in crystalline **chlorine, bromine** and **iodine** are packed in a different manner, as shown in Fig. 43. The rather different distances between atoms of adjacent molecules are remarkable. If we take the VAN DER WAALS distance, such as observed in organic and inorganic molecular compounds, as reference, then some of the intermolecular contacts in the b-c plane are shorter, whereas they are longer to the molecules of the next plane. We thus observe a certain degree of association of the halogen molecules within the b-c plane. This association increases from chlorine to iodine. The weaker attractive forces between the planes show up in the plate-like habit of the crystals and in their easy cleavage parallel to the layers. Similar association tendencies are also observed for the heavier elements of the fifth and sixth main group; they are discussed in the following sections.

The packing can be interpreted as a cubic-closest packing of halogen atoms that has been severely distorted by the covalent bonds within the molecules.

---

*For literature about the structures of all elements, see [59].

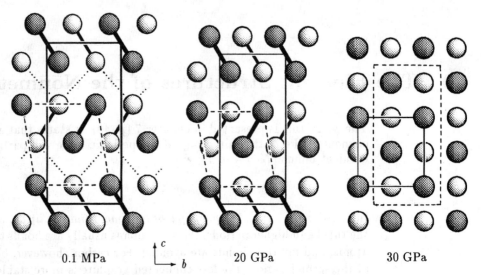

Fig. 43
The structure of
iodine at three
different
pressures. The
outlined
face-centered
unit cell in the
right figure
corresponds to
that of a
(distorted)
cubic closest
sphere packing

0.1 MPa   20 GPa   30 GPa

By exerting pressure the distortion is reduced, i.e. the different lenghts of the
contact distances between the atoms approximate each other (Fig. 43). For iodine a continuous approximation is observed with increasing pressure, then at
21 GPa an abrupt phase transition takes place which ends in a nearly undistorted cubic closest-packing of spheres. With increasing pressure the energy
gap between the fully occupied valence band and the unoccupied conduction
band decreases and finally disappears at about 18 GPa, i.e. a transition from
an insulator to a metallic conductor takes place. Iodine thus not only adopts
the structure of a (slightly distorted) closest sphere packing that is typical for
metals, but actually becomes a metal. A comparable transition from a molecular structure to a metal is also believed to occur for hydrogen; the necessary
pressure (not yet achievable experimentally) is estimated to be from 20 to 60
TPa (2 to 6×10^{13} Pa).

10.2   Chalcogens

Oxygen in the solid state also consists of O_2 molecules. Below 43.6 K they are
packed as in α-F_2, and above 43.6 K as in β-F_2.

No element shows as many different structures as **sulfur**. Crystal structures
are known for the following forms: S_6, S_7 (two modifications), S_8 (three mod-
ifications), S_{10}, $S_6 \cdot S_{10}$, S_{11}, S_{12}, S_{13}, S_{18} (two forms), S_{20}, S_∞. Many of them
can be separated by chromatography from solutions that were obtained by ex-
traction of quenched sulfur melts. Further forms with unestablished structures
exist. For instance, by quenching sulfur vapor from 770 K to 60 K violet sulfur
is obtained; its paramagnetism indicates the possible presence of S_2 molecules;
when heated to 80 K it becomes green. Quenched sulfur melts also yield polymer forms; the structure of one of these has been determined. All structurally
characterized sulfur forms consist of rings or chains of S atoms, every sulfur

atom being bonded with two other sulfur atoms in the sense of the $8 - N$ rule. The S–S bond lengths usually are about 206 pm, but they show a certain scatter (± 10 pm). The S–S–S bond angles are between 101 and 110° and the dihedral angles between 74 and 100°.* As a consequence, a sequence of five atoms can adopt one of two arrangements:

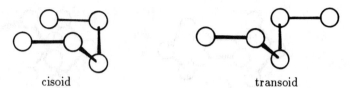

cisoid   transoid

In the smaller rings S_6, S_7 and S_8 only the cisoid arrangement occurs, the dihedral angles being forced to adapt themselves (74.5° for S_6, 98° for S_8). S_6 has chair conformation, and the S_8 conformation is called crown-form (Fig. 44). S_7 can be imagined to be formed from S_8 by taking out one S atom. Larger rings require the presence of cisoid and transoid groups in order to be free of strain. In S_{12} cisoid and transoid groups alternate. Helical chains result when there are only transoid groups; the dihedral angle determines how many turns it takes to reach another atom directly above of an atom on a line parallel to the axis of the helix. In the only structurally well-characterized form of polymer sulfur it takes ten atoms in three turns.†

In othorhombic α-sulfur, the modification stable at normal conditions, S_8 rings are stacked to form columns. Consecutive rings are not stacked one exactly above another (as in a roll of coins), but in a staggerd manner so that the column looks like a crank shaft (Fig. 45). This arrangement allows for a dense packing of the molecules, with columns in two mutually perpendicular directions. The columns of one direction are placed in the recesses of the perpendicular "crank shafts". In S_6 and in S_{12} the rings are stacked exactly one above another, and the rolls are bundled parallel to each other. In the structures a generally valid principle can be discerned: *in the solid state molecules tend to pack as tightly as possible.*

Selenium forms three known modifications that consist of Se_8 rings. The stacking of the rings differs from that of the S_8 modifications in that they resemble coin rolls, but the rings are tilted. The thermodynamically stable form of selenium, α-selenium, consists of helical chains having three Se atoms in every turn (Fig. 46). The chains are bundled parallel in the crystal. Every selenium atom has four adjacent atoms from three different chains at a distance of 344 pm. Together with the two adjacent atoms within the chain at a distance of 237 pm, a strongly distorted octahedral 2+4 coordination results. The Se$\cdots$Se distance between the chains is significantly shorter than expected from the VAN DER WAALS distance.

---

*dihedral angle = angle between two planes. For a chain of four atoms it is the angle between the planes through the atoms 1,2,3 and 2,3,4.

†The helix has the symmetry of a 10_3 screw axis (cf. section 18.1).

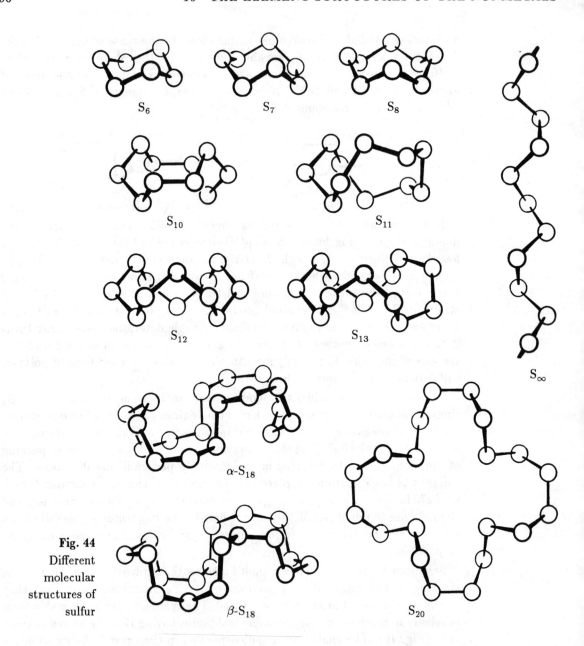

Fig. 44
Different
molecular
structures of
sulfur

Tellurium crystallizes isotypic to α-selenium. As expected, the Te–Te bonds in the chain (283 pm) are longer than in selenium, but the contact distances to the atoms of the adjacent chains are nearly the same (Te$\cdots$Te 349 pm). The shortening, as compared to the VAN DER WAALS distance, is more marked and the deviation from a regular octahedral coordination of the atoms is reduced (cf. Table 14, p. 100). By exerting pressure all six distances can be made to be equal; above 7 GPa every tellurium atom has six equidistant adjacent atoms at a distance of 300 pm, and the structure corresponds to that of β-polonium.

Fig. 45
Section of the
structure of
α-sulfur

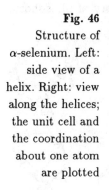

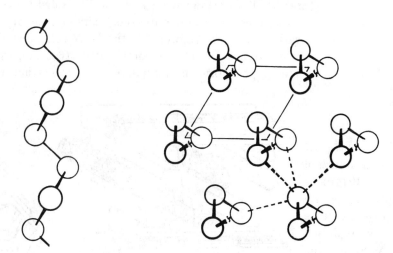

Fig. 46
Structure of
α-selenium. Left:
side view of a
helix. Right: view
along the helices;
the unit cell and
the coordination
about one atom
are plotted

Two modifications are known for **polonium**. At room temperature α-polonium is stable; it has a cubic-primitive lattice, every atom having an exact octahedral coordination (Fig. 4, p. 8). This is a rather unusual structure for a metal, but it also occurs for phosphorus and antimony at high pressures. At 54 °C α-Po is converted to β-Po. During the phase transition the lattice suffers a compression in the direction of one of the body diagonals of the cubic-primitive lattice, and the result is a rhombohedral lattice with an angle of 98.1°.

10.3   Elements of the Fifth Main Group

For solid **nitrogen** three modifications are known that differ in the packing of the N_2 molecules. Two of them are stable at normal pressure (transition temperature 35.6 K); the third one exists only under high pressure.

Phosphorus exhibits several modifications, some of which have been structurally characterized, while the structures of a larger number of forms of red phosphorus remain obscure. Phosphorus vapor consists of tetrahedral P_4 molecules, and at higher temperatures also of P_2 molecules ($P\equiv P$ distance 190 pm). White phosphorus forms by condensation of the vapor. It also consists, like liquid phosphorus, of P_4 molecules.

By irradiation with light or by heating it to temperatures above 180 °C, white phosphorus is transformed to red phosphorus. Its tint, melting point, vapor pressure and especially its density depend on the conditions of preparation. It is amorphous or microcrystalline, and several different modifications with unknown polymer structures appear to exist.

HITTORF's (violet) phosphorus forms by slow crystallization at temperatures around 550 °C; single crystals were obtained by slow cooling (from 630 to 520 °C) of a solution in liquid lead. It has a rather complicated structure, in which cages of the same shape as in As_4S_4 and As_4S_5 are connected via further P atoms in such a way that five-sided tubes result (Fig. 47). The tubes are connected crosswise to grids; pairs of grids are interlocked but not bonded with each other. According to the $8-N$ rule every P atom is bonded with three other atoms. Despite its complicated structure, the linking principle of HITTORF's phosporus occurs frequently in the structural chemistry of phosphorus

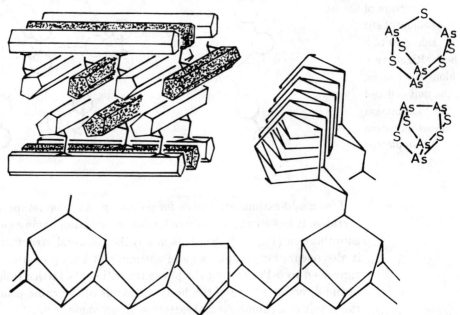

Fig. 47
Constitution of
HITTORF's
phosphorus with
cages like those
in As_4S_4 and
As_4S_5. Bottom:
two joined
five-sided tubes.
Upper left: two
interlocked grids
of such tubes;
hatched and
white tubes are
not connected
with each other

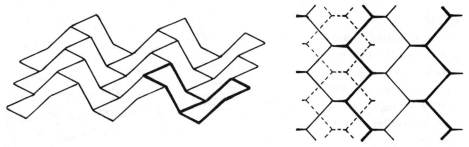

Fig. 48
The structure of
black phosphorus.

Left: section of one layer; two rings with chair conformation and relative arrangement as in *cis*-decaline are emphasized. Right: top view of a layer showing the zigzag lines; the position of the next layer is indicated

compounds; building units that correspond to fragments of the tubes are known among the polyphosphides and polyphosphanes (cf. p. 122).

Black phosphorus only forms under special conditions (high pressure, crystallization from liquid Bi or prolonged heating in the presence of Hg); nevertheless, it is the thermodynamically stable modification at normal conditions. It consists of layers having six-membered rings in the chair conformation. Pairs of rings are connected like the rings in *cis*-decaline (Fig. 48). The layer can also be regarded as a system of interconnected zigzag lines that alternate in two different planes. Within the layer every P atom is bonded to three other P atoms at distances of 222 and 224 pm. Somewhat further away there are another two atoms from the next but one zigzag line at a distance of 331 pm, and two more atoms at 359 pm and 380 pm from the next layer. The distances between the layers correspond to the VAN DER WAALS distance. Under pressure black phosphorus is transformed to a modification with the structure of arsenic, and this converts at even higher pressures to the α-polonium structure. Certain structural features of black phosphorus are also found among the polyphosphides (cf. Fig. 64, p. 122).

Arsenic modifications with the structures of white and black phosphorus have been described. However, only gray (metallic, rhombohedral) α-arsenic is stable. It consists of layers of six-membered rings in the chair conformation that are connected with each other in the same way as in *trans*-decaline (Fig. 49). In the layer the atoms are situated alternately in an upper and a lower plane. The layers are stacked in a staggered manner such that over and under the center of every ring there is an As atom in an adjacent layer. This way every As atom is in contact with three more atoms in addition to the three atoms to which it is bonded within the layer; it has a distorted octahedral 3+3 coordination. The As–As bond length in the layer is 252 pm; the distance between adjacent atoms of different layers is 312 pm and thus is considerably shorter than the VAN DER WAALS distance (370 pm).

The structures of **antimony** and **bismuth** correspond to that of gray arsenic. With increasing atomic weight the distances between adjacent atoms within a

Fig. 49

Section of a layer in gray arsenic and the position of two rings of the next layer. Two rings with the relative arrangement as in *trans*-decaline are emphasized

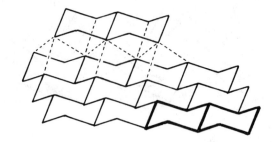

layer and between layers become less different, i.e. the coordination polyhedra deviate less from a regular octahedron. This effect is enhanced under pressure (Table 14); it corresponds to the observations for selenium and tellurium.

At even higher pressures antimony and bismuth adopt typical metal structures: at 9 GPa Sb has a hexagonal closest sphere packing and Bi a body-centered cubic sphere packing.

The description of the structures of P, As, Sb, and Bi as layer structures and of Se and Te as chain structures neglects the presence of bonding interactions between the layers and chains, respectively. These interactions gain importance for the heavier atoms. For instance, the interlayer distances between adjacent atoms for Sb and Bi are only 15 % longer than the intralayer distances; the actual deviation from the α-polonium structure is rather small. Furthermore, As, Sb, and Bi show metallic conductivity. The bonding interactions can be understood using band theory: starting from the α-Po structure, a PEIERLS distortion takes place that enhances three bonds per atom. Since the $6p$ band of polonium intersects with other unoccupied bands, the $6p$ band remains unoccupied at higher energy levels, and the distortion is avoided for polonium. In the series bismuth, antimony, arsenic the possibilities for the electrons to switch

Table 14 Distances between adjacent atoms and bond angles in structures of the α-As, α-Se, α-Po and β-Po type. d_1 = bond distance, d_2 = shortest interatomic distance between layers or chains, respectively; distances in pm, angles in degrees

	structure type	d_1	d_2	d_2/d_1	angle
P (~10 GPa)	α-As	213	327	1.54	105
P (~12 GPa)	α-Po	238	238	1.00	90
As	α-As	252	312	1.24	96.6
Se	α-Se	237	344	1.45	103.1
Sb	α-As	291	336	1.15	95.6
Sb (5 GPa)	α-Po	299	299	1.00	90
Te	α-Se	283	349	1.23	103.2
Te (11.5 GPa)	β-Po	300	300	1.00	103.3
Bi	α-As	307	353	1.15	95.5
Po	α-Po	337	337	1.00	90
Po	β-Po	337	337	1.00	98.1

over to other bands decreases, and the distortion becomes more marked. The same applies for tellurium and selenium. Fig. 50 shows how the α-As and α-Se structures result by PEIERLS distortion from the α-polonium structure.

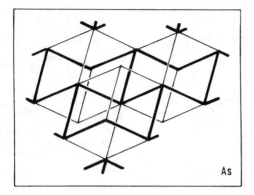
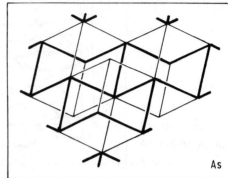
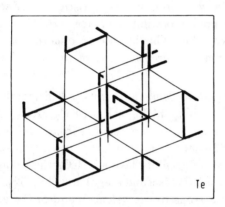
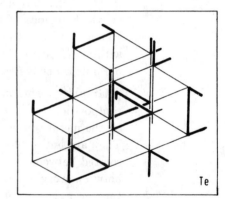

Fig. 50
The layer and chain structures of the elements of the fifth and sixth main group result by elongation of certain distances in the α-Po structure (stereo images)

10.4   Carbon

Graphite is the modification of carbon which is stable under normal conditions. It has a structure consisting of planar layers (Fig. 51). Within the layer each C atom is bonded covalently with three other C atoms. Every atom contributes one p orbital and one electron to the delocalized π bond system of the layer. This constitutes a half-filled band, so we have a two-dimensional metallic state with the corresponding electrical conductivity. Between the layers weak VAN DER WAALS forces are the essential attractive forces. The bonds within the layers have a length of 142 pm and the distance from layer to layer is 335 pm. The high electric conductivity therefore only exists parallel to the layers, but not perpendicular to them. The layers are stacked in a staggered manner; half of the atoms of one layer are situated exactly above atoms of the layer below, and the other half are situated over the ring centers (Fig. 51). Three layer positions are possible, *A, B* and *C*. The stacking sequence in normal (hexagonal) graphite

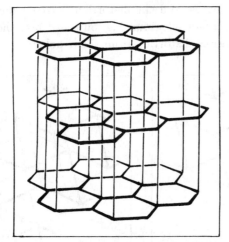 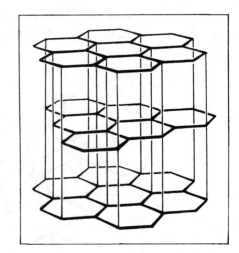

Fig. 51
Structure of
graphite (stereo
image)

is $ABAB\ldots$, but frequently a more or less statistical layer sequence is found, in which regions of the predominating sequence $ABAB\ldots$ are separated by regions with the sequence ABC. This is called a *one-dimensional disorder*, i.e. within the layers the atoms are ordered, but in the direction of stacking the periodic order is missing.

Graphite forms *intercalation compounds* with alkali metals [91]. They have compositions such as LiC_6, LiC_{12}, LiC_{18} or KC_8, KC_{24}, KC_{36}, KC_{48}. Depending on the metal content they have colors extending from a golden luster to black. They are better electrical conductors than graphite. The alkali ions are intercalated between every pair of graphite layers in KC_8 (first stage intercalation), between every other pair in KC_{24} (second stage) etc. (Fig. 52). The metal atoms turn over their valence electrons to the valence band of the graphite. A different kind of intercalation compound are those with metal chlorides MCl_n (M = nearly all metals; $n = 2$ to 6) and some fluorides and bromides. The intercalated halide layers have structures that essentially correspond to the structures in the pure compounds; for example, intercalated $FeCl_3$ layers have the same structure as in pure $FeCl_3$, as shown on the front cover.

Carbon in its different forms such as pit-coal, coke, charcoal, soot etc., is in principle graphite-like, but with a low degree of ordering. It can be microcrystalline or amorphous; OH groups and possibly other atom groups are bonded at the edges of the graphite layer fragments. Many species of carbon have numerous pores and therefore have a large inner surface; for this reason they can adsorb large quantities of other substances and act as catalysts. In this respect crystalline graphite is less active. Carbon fibers also consist of graphite layers that are oriented parallel to the fiber direction. By deposition from the gas phase, carbon fibers can be made that consist of tubes having concentric, closed, bent graphite layers that are piled one around the other; as the curvature cannot be smaller than some limiting value, the tubes remain hollow.

Fig. 52
Left: arrangement of the K^+ ions relative to an adjacent graphite layer in KC_8; in KC_{24} a K^+ ion layer only contains two thirds as many ions, they are disordered and highly mobile. Right: stacking sequence of graphite layers and K^+ ions in KC_8 and KC_{24}

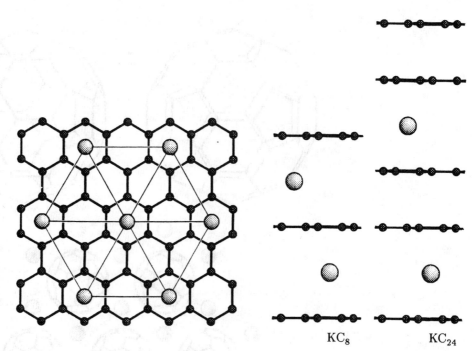

KC_8 $\qquad$ KC_{24}

Fullerenes are modifications of carbon that consist of cagelike molecules [92]. They can be obtained by setting up an electric arc between two graphite electrodes in a controlled atmosphere of helium and condensing the evaporated carbon, and then recrystallizing it from magenta-colored benzene solution. The main product is the fullerene C_{60}, called buckminsterfullerene* (buckyball for short). The C_{60} molecule has the shape of a soccer ball, consisting of 12 pentagons and 20 benzene-like hexagons (Fig. 53). Second in yield from this preparation is C_{70}, which has 12 pentagons and 25 hexagons and a shape reminiscent of a peanut. Cages with other sizes can also be produced, but they are less stable (they may have any even number of C atoms, beginning at C_{32}).

In crystalline C_{60} the molecules have a face-centered cubic arrangement, i.e. they are packed as in a cubic closest-packing of spheres; as they are nearly spherical, the molecules spin in the crystal. The crystals are as soft as graphite. Similar to the intercalation compounds of graphite, potassium atoms can be enclosed; they occupy the cavities between the C_{60} balls. With all cavities occupied (tetrahedral and octahedral interstices if the C_{60} balls are taken as closest-packed spheres), the composition is K_3C_{60}. This compound has metallic properties and becomes superconducting when cooled below 18 K.

The structures of diamond, silicon, germanium and tin are discussed in more detail in chapter 11.

---

*Buckminsterfullerene is named after the engineer Buckminster Fuller who invented the geodesic dome, which works on the same architectural principle as the C_{60} molecule.

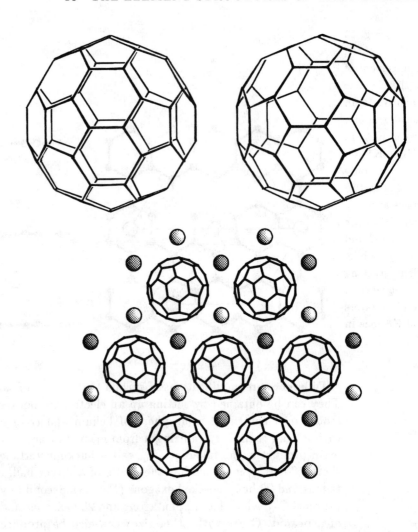

Fig. 53
Top: molecular structure of the C_{60} molecule (stereo view). Bottom: packing of C_{60} molecules and K^+ ions in K_3C_{60}

10.5   Boron

Boron is as unusual in its structures as it is in its chemical behavior. Sixteen boron modifications have been described, but most of them have not been well characterized. Many samples assumed to have consisted only of boron were possibly boron-rich borides (many of which are known, e.g. YB_{66}). An established structure is that of rhombohedral $\alpha\text{-}B_{12}$ (the subscript number designates the number of atoms per unit cell). The crystal structures of two further forms are known, rhombohedral $\beta\text{-}B_{105}$ and tetragonal $\alpha\text{-}B_{50}$, but in both cases probably boron-rich borides were studied. $\alpha\text{-}B_{50}$ should be formulated $B_{48}X_2$. It consists of B_{12} icosahedra that are linked by tetrahedrally coordinated X atoms. These atoms are presumably C or N atoms (B, C and N can hardly be distinguished by X-ray diffraction).

The outstanding building unit in all modifications of boron that have been described is the B_{12} icosahedron, which also is present in the anionic *closo*-borane

$B_{12}H_{12}^{2-}$. The twelve atoms of an icosahedron are held together by multicenter bonds; according to MO theory, 13 bonding orbitals occupied by 26 electrons should be present; 10 valence electrons are left over. In the $B_{12}H_{12}^{2-}$ ion 14 additional electrons are present (12 from the H atoms, 2 from the ionic charge), which amounts to a total of 24 electrons or 12 electron pairs; these are used for the 12 covalent B–H bonds that are oriented radially outwards from the icosahedron. In elemental boron the B_{12}-icosahedra are linked with one another by such radial bonds, but for 12 bonds only 10 valence electrons are available; therefore, not all of them can be normal two-electron two-center bonds.

In α-B_{12} the icosahedra are arranged as in a cubic closest-packing of spheres (Fig. 54). In one layer of icosahedra every icosahedron is surrounded by six other icosahedra that are linked by two-electron three-center bonds. Every boron atom involved contributes an average of $\frac{2}{3}$ electrons to these bonds, which amounts to $\frac{2}{3} \cdot 6 = 4$ electrons per icosahedron. Every icosahedron is surrounded additionally by six icosahedra of the two adjacent layers, to which it is bonded by normal B–B bonds; this requires 6 electrons per icosahedron. In total, this adds up exactly to the above-mentioned 10 electrons for the inter-icosahedron bonds.

Fig. 54
Structure of rhombohedral α-B_{12}. The icosahedra in the layer section shown are connected with each other by 2e3c-bonds. One icosahedron of the next layer is shown

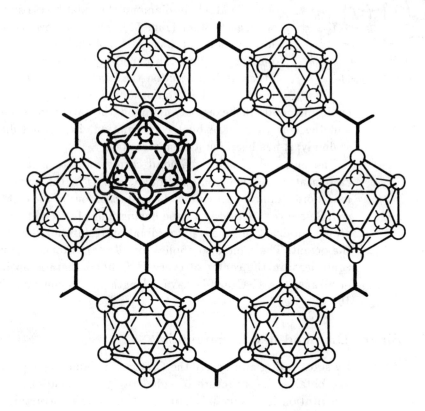

11 Diamond-like Structures

11.1 Cubic and Hexagonal Diamond

Diamond, silicon, germanium and (gray) α-tin (stable below 13 °C) are isotypic. Diamond consists of a network of carbon atoms with four covalent bonds per atom. Starting from a layer of gray arsenic (cf. Fig. 49), all As atoms can be thought of as being substituted by C atoms; each of these can participate in a fourth bond that is oriented perpendicular to the layer. Relative to any one of the chair conformation rings of the layer the bonds within the layer take equatorial positions; the remaining bonds correspond to axial positions that are directed alternately upwards and downwards from the layer. In graphite fluoride $(CF)_x$ every axial position is occupied by a fluorine atom. In diamond the axial bonds serve to link the layers with each other (Fig. 55). Thereby new six-membered rings are formed that can have either a chair or a boat conformation, depending on how the joined layers are positioned relative to each other. If in projection the layers are staggered, then all resulting rings have chair conformation; this is the arrangement in normal, cubic diamond. In hexagonal diamond the layers in projection are eclipsed, and the new rings have boat conformation. Hexagonal diamond occurs very seldomly; it has been found in meteorites.

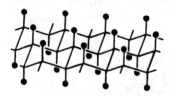

graphite fluoride

The unit cell of cubic diamond corresponds to a face-centered packing of carbon atoms. Aside from the four C atoms in the vertices and face centers, four more atoms are present in the centers of four of the eight octants of the unit cell. Since every octant is a cube having four of its eight vertices occupied by C atoms, an exact tetrahedral coordination results for the atom in the center of the octant. The same also applies for all other atoms—they are all symmetry-equivalent; in the center of every C–C bond there is an inversion center. As in alkanes the C–C bonds have a length of 154 pm and the bond angles are 109.47°.

11.2 Binary Diamond-like Compounds

By substituting alternately the C atoms in cubic diamond by Zn and S atoms, one obtains the structure of sphalerite (zinc blende). By the corresponding substitution in hexagonal diamond, the wurtzite structure results. As long as atoms of one element are allowed to be bonded only to atoms of the other element, binary compounds can only have a 1:1 composition. For the four bonds per atom an average of four electrons per atom are needed; this condition is

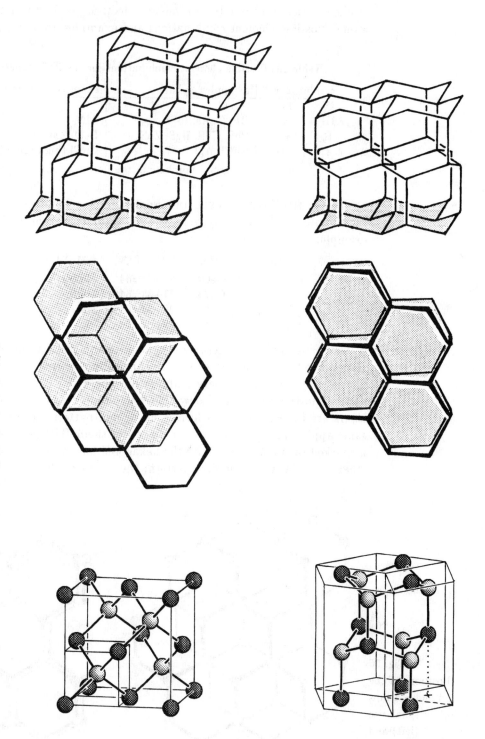

Fig. 55
Structure of cubic (left) and hexagonal (right) diamond. Top row: connected layers as in α-As. Central row: the same layers in projection perpendicular to the layers. Bottom: unit cells; when the light and dark atoms are different, this corresponds to the structures of sphalerite (zinc blende) and wurtzite, respectively

fulfilled if the total number of valence electrons is four times the number of atoms. Possible element combinations and examples are given in Table 15.

Table 15 Possible element combinations for the ZnS structure types

combination*		examples, sphalerite-type	examples, wurtzite-type
IV	IV	β-SiC	
III	V	BP, GaAs, InSb	AlN
II	VI	BeS, CdS, HgS, ZnSe	BeO, ZnO, CdS (high temp.)
I	VII	CuCl, AgI	CuCl (high temp.)

* group numbers in the periodic table

The GRIMM–SOMMERFELD rule applies for the bond lengths: if the sum of the atomic numbers is the same, the interatomic distances are the same. For example:

MX	$Z(M)+Z(X)$	$d(M-X)$
GeGe	32+32=64	245.0 pm
GaAs	31+33=64	244.8
ZnSe	30+34=64	244.7
CuBr	29+35=64	246.0

The sections of the structures of sphalerite and wurtzite shown in Fig. 56 correspond to the central row of Fig. 55 (projections perpendicular to the arsenic-like layers). Behind of every sulfur atom there is a zinc atom bonded to it situated in the direction of view. The zinc atoms within one of the arsenic-like layers are in one plane and form a hexagonal pattern (dotted in Fig. 56); the same applies for the sulfur atoms on top of them. The position of the pattern is marked by an *A*. In wurtzite the hexagonal pattern of the following atoms is staggered relative to the first pattern; the atoms of this position *B* are placed

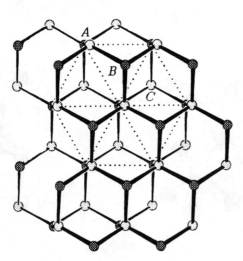
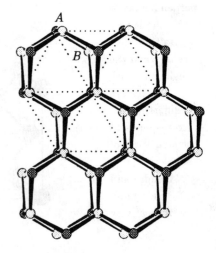

Fig. 56
Positions of the
Zn and S atoms
in sphalerite
(left) and
wurtzite

over the centers of one half of the dotted triangles. Atoms over the centers of the remaining triangles (position C) do not occur in wurtzite, but they do occur in sphalerite. If we designate the positions of the planes containing the Zn atoms by A, B, and C, respectively, and the corresponding planes of the S atoms by α, β, and γ, then the following stacking sequences apply for the planes:

sphalerite: $A\alpha B\beta C\gamma \ldots$ wurtzite: $A\alpha B\beta \ldots$

Other stacking sequences than these are also possible, for example $A\alpha B\beta A\alpha C\gamma \ldots$ or statistical sequences without periodic order. More than 70 stacking varieties are known for silicon carbide, and together they are called α-SiC. Structures that can be considered as stacking variants are called polytypes. We deal with them further in the context of closest sphere packings (chapter 13).

Several of the binary diamond-like compounds have industrial applications because of their physical properties. They include silicon carbide and cubic boron nitride (obtainable from graphite-like BN under pressure at 1800 °C); they are almost as hard as diamond and serve as abrasives. SiC is also used to make heating devices for high temperature furnaces as it is a semiconductor with a sufficiently high conductivity at high temperatures, but also is highly corrosion resistant and has a low thermal expansivity. Yellow CdS and red CdSe are used as color pigments, and ZnS as a luminophore in cathode ray displays. The III–V compounds are semiconductors with electric properties that can be adapted by variation of the composition and by doping; light-emitting diodes and photovoltaic cells are made on a base of GaAs.

11.3 Diamond-like Compounds under Pressure

The diamond-type structure of α-tin is stable at ambient pressure only up to 13 °C; above 13 °C it transforms to β-tin (white tin). The transition α-Sn $\rightarrow$ β-Sn can also by achieved below 13 °C by exerting pressure. Silicon and germanium also adopt the structure of β-Sn at higher pressures (over 1.2 GPa, approximately). The transformation involves a considerable increase in density (for Sn +21%). The β-Sn structure evolves from the α-Sn structure by a drastic compression in the direction of one of the edges of the unit cell (Fig. 57). This way two atoms that previously were further away in the direction of the compression become neighbors to an atom; together with the four atoms that were already adjacent in α-Sn, a coordination number of 6 results. The regular coordination tetrahedron of α-Sn is converted to a flattened tetrahedron with Sn–Sn distances of 302 pm; the two atoms above and below the flattened tetrahedron are at a distance of 318 pm. These distances are *longer* than in α-Sn (281 pm). Although β-Sn forms from α-Sn by the action of pressure and has a higher density, the transformation involves an increase of the interatomic distances.

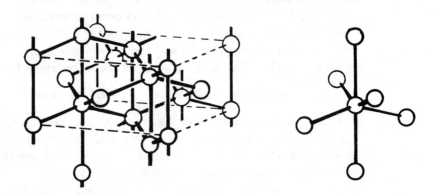

Fig. 57

Structure of white tin (β-Sn). The dashed cell corresponds to a unit cell of diamond (α-Sn) that has been strongly compressed in one direction. Right: coordination about an Sn atom

Generally, the following rules apply for pressure-induced phase transitions:

Pressure–coordination rule by A. NEUHAUS: *with increasing pressure an increase of the coordination number takes place.*

"Pressure–distance paradox" by W. KLEBER: *When the coordination number increases according to the previous rule, the interatomic distances also increase.*

Further examples where these rules are observed are as follows. Under pressure, some compounds with sphalerite structure such as AlSb and GaSb, transform to modifications that correspond to the β-Sn structure. Others such as InAs, CdS, and CdSe, adopt the NaCl structure when compressed, and their atoms thus also attain coordination number 6. Graphite (c.n. 3, C–C distance 141.5 pm, density 2.26 g cm^{-3}) $\xrightarrow{\text{pressure}}$ diamond (c.n. 4, C–C 154 pm, 3.51 g cm^{-3}).

11.4 Polynary Diamond-like Compounds

Of the numerous ternary and polynary diamond-like compounds we deal only with those that can be considered as superstructures of sphalerite. A superstructure is a structure that, while having the same structural principle, has an enlarged unit cell. When the unit cell of sphalerite is doubled in one direction (c axis), different kinds of atoms can occupy the doubled number of atomic positions. All the structure types listed in Fig. 58 have the tetrahedral coordination of all atoms in common, except for the variants with certain vacant positions.

CuFeS$_2$ (chalcopyrite) is one of the most important copper minerals. Red β-Cu$_2$HgI$_4$ and yellow β-Ag$_2$HgI$_4$ (CdGa$_2$S$_4$ type) are thermochromic: they transform at 70 °C and 51 °C, respectively, to modifications having different colors (black and orange, respectively); in these the atoms and the vacancies have a disordered distribution.

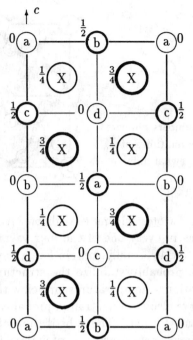

	Atomic position				
structure type	X	a	b	c	d
CuFeS₂*	S	Fe	Cu	Fe	Cu
Cu₃SbS₄[†]	S	Sb	Cu	Cu	Cu
Cu₂FeSnS₄[‡]	S	Fe	Sn	Cu	Cu
CdGa₂S₄	S	Cd	Ga	Ga	□
β-Cu₂HgI₄	I	Hg	□	Cu	Cu

□ = vacancy
* chalcopyrite
[†] famatinite
[‡] stannite

The numbers next to the circles designate the height in the direction of view

Fig. 58
Superstructures
of the sphalerite
type with
doubled *c* axis

Aside from the superstructures mentioned, other superstructures with other enlargement factors for the unit cell are known, as well as superstructures of wurtzite. Defect structures, i.e. structures with vacancies, are known with ordered and disordered distributions of the vacancies. γ-Ga₂S₃, for instance, has sphalerite structure with statistically only two thirds of the metal positions occupied by Ga atoms.

11.5   Widened Diamond Lattices. SiO₂ Structures

Take elemental silicon (diamond structure) and insert an oxygen atom between every pair of silicon atoms; this way, every Si–Si bond is replaced by an Si–O–Si group and every Si atom is surrounded tetrahedrally by four O atoms. The result is the structure of cristobalite. The SiO₄ tetrahedra are all linked by common vertices. As there are twice as many Si–Si bonds than Si atoms in silicon, the composition is SiO₂. Cristobalite is one of the polymorphous forms of SiO₂; it is stable between 1470 and 1713 °C and is metastable at lower temperatures. It occurs as a mineral. The oxygen atoms are situated to the side of the Si$\cdots$Si connecting lines, so that the Si–O–Si bond angle is 147°. The structure model shown in Fig. 59, however, is only a snapshot. Above 250 °C the tetrahedra perform coupled tilting vibrations that on average feign a higher symmetry, with O atoms exactly on the Si–Si connecting lines. When cooled below ~240 °C the vibrations "freeze" ($\rightarrow$ α-cristobalite; the $\alpha \rightleftharpoons \beta$ transition temperature depends on the purity of the sample).

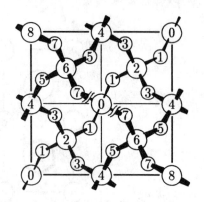

Fig. 59

Snapshot of the unit cell of β-cristobalite. The numbers indicate the height of the atoms in the direction of view as multiples of $\frac{1}{8}$. Below 240 °C the SiO$_4$ tetrahedra are tilted in a somewhat different manner (α-cristobalite)

The insertion of the oxygen atoms widens the silicon lattice considerably; a relatively large void remains in every one of the four vacant octants of the unit cell. In natural cristobalite they usually contain foreign ions (mainly alkali and alkaline earth ions) that probably stabilize the structure and allow the crystallization of this modification at temperatures far below the stability range of pure cristobalite. To conserve electrical neutrality, probably one Si atom per alkali ion is substituted by an Al atom.[*] The substitution of Si by Al atoms in an SiO$_2$ framework with simultaneous inclusion of cations in voids is a very common phenomenon; silicates of this kind are called aluminosilicates. The mineral carnegieite Na[AlSiO$_4$] has a cristobalite structure in which one half of the Si atoms have been substituted by Al atoms and all voids have been occupied by Na$^+$ ions. The LOEWENSTEIN rule has been stated for aluminosilicates: AlO$_4$ tetrahedra tend not to be linked directly with each other; the group Al–O–Al is avoided.

Tridymite is another form of SiO$_2$ which is stable between 870 and 1470 °C, but it can also be maintained in a metastable state at lower temperatures and occurs as a mineral. Its structure can be derived from that of hexagonal diamond in the same way as that of cristobalite from cubic diamond. In this case the oxygen atoms are also situated to the side of the Si$\cdots$Si connecting lines and the Si–O–Si bond angles are approximately 150°. At temperatures below 380°C several variants occur that differ in the kind of the mutual tilting of the SiO$_4$ tetrahedra. Tridymite also encloses larger voids that can be occupied by alkali or alkaline earth ions. The anionic framework of some aluminosilicates corresponds to the tridymite structure, for example in nepheline, Na$_3$K[AlSiO$_4$]$_4$.

Quartz is the modification of SiO$_2$ that is stable up to 870 °C, with two slightly different forms, α-quartz occurring below and β-quartz above 573 °C. The transition involves merely a minor mutual rotation of the SiO$_4$ tetrahedra. We discuss the quartz structure here, although it cannot be derived from one of the forms of diamond. Nevertheless, quartz also consists of a network of SiO$_4$ tetrahedra sharing vertices, but with smaller voids than in cristobalite

---

[*]Al and Si can hardly be distinguished by X-ray structure analysis due to the nearly equal number of electrons

Fig. 60

The structure of α-quartz. Only SiO₄ tetrahedra are shown. Numbers designate the heights of the Si atoms in the tetrahedron centers as multiples of $\frac{1}{3}$ of the unit cell height. The symbols ▲ mark the axes of two helical chains (stereo image)

and tridymite (this is manifested in the densities: quartz 2.65, cristobalite 2.33, tridymite 2.27 g cm⁻³). As shown in Fig 60, the tetrahedra form helices, and in a given crystal these are all either right-handed or left-handed. Right-handed and left-handed quartz can also be intergrown in a well-defined manner, forming twinned crystals ("Brazilian twins"). Due to the helical structure quartz crystals are optically active and have piezoelectric properties (section 17.2). Quartz crystals are produced industrially by hydrothermal synthesis. For this purpose quartz powder is placed in one end of a closed vial at 400 °C, while a seed crystal is placed at the opposite end at 380 °C; the vial is filled with an alkaline aqueous solution that is maintained liquid by a pressure of 100 MPa. The quartz powder slowly dissolves while the seed crystal grows.

The phase diagram for SiO₂ is shown in Fig. 61. The transitions between the α and β forms only require minor rotations of the SiO₄ tetrahedra, the linkage pattern remaining unaltered; these transitions take place rapidly. The other transitions, on the other hand, require a reconstruction of the structure, Si–O bonds being untied and rejoined; they proceed slowly and thus render possible the existence of the metastable modifications. In addition to the mentioned modifications other polymorphs are known that form under pressure, namely coesite, keatite, and stishovite. Coesite and keatite also consist of frameworks of SiO₄ tetrahedra sharing vertices. Stishovite, however, has the rutile structure, i.e. silicon atoms having coordination number 6.

Further compounds that occur with the structure types of SiO₂ are H₂O and BeF₂. Ice normally crystallizes in the hexagonal tridymite type (ice I_h), the oxygen atoms occupying the Si positions of tridymite while the hydrogen atoms

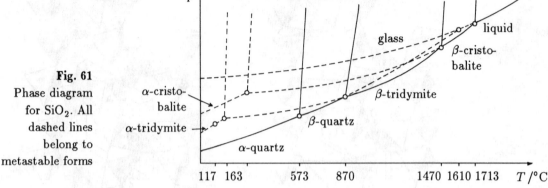

Fig. 61
Phase diagram
for SiO$_2$. All
dashed lines
belong to
metastable forms

are placed between two oxygen atoms each. An H atom is shifted towards one of the O atoms so that it belongs to one H$_2$O molecule and participates in a hydrogen bridge to another H$_2$O molecule. Ice I$_c$ crystallizes from the gas phase at temperatures below $-140\,°$C; it has a cubic structure like cristobalite. Under pressure nine further modifications can be obtained, some of which correspond to other SiO$_2$ modifications (e.g. keatite; see the phase diagram of H$_2$O, Fig. 8, p. 16).

In the same way as the sphalerite structure is derived from diamond by the alternating substitution of C atoms by Zn and S atoms, the Si atoms in the SiO$_2$ structures can be substituted alternately by two different kinds of atoms. Examples include AlPO$_4$, MnPO$_4$, and ZnSO$_4$. The cristobalite and tridymite structures with filled voids also are frequently encountered. Examples in addition to the above-mentioned aluminosilicates Na[AlSiO$_4$] and Na$_3$K[AlSiO$_4$]$_4$ are K[FeO$_2$] and MILLON's base, [NHg$_2$]$^+$OH$^-\cdot$H$_2$O.

The large voids in the network of cristobalite can also be filled in another way, namely by a second identical network that interpenetrates the first. The structure of cuprite, Cu$_2$O, has this structure. Take a cristobalite structure in which the Si positions are occupied by O atoms that are linked via Cu atoms having coordination number 2. As the bond angle at a Cu atom is 180°, the packing density is even less than in cristobalite itself. Two exactly equal networks of this kind interpenetrate each other, one being shifted against the other (Fig. 62). The two networks "float" one within the other; there are no direct bonds between them. This kind of a structure is possible when tetrahedrally coordinated atoms are held at a distance from each other by linear linking groups like –Cu– or –Ag– (in isotypic Ag$_2$O). Cyanide groups between tetrahedrally coordinated zinc atoms, Zn–C≡N–Zn, act in the same way as spacers in Zn(CN)$_2$, which has the same structure as Cu$_2$O (with metal atom positions interchanged with the anion positions).

Fig. 62
The structure of
Cu$_2$O (cuprite).
Eight unit cells
are shown; they
correspond to
one unit cell of
cristobalite. The
gray network has
no direct bonds
to the black
network (stereo
image)

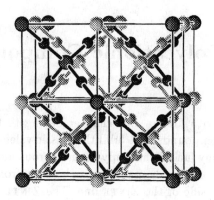 

11.6 Problems

11.1 The bond length in β-SiC is 188 pm. For which of the following compounds would you expect longer, shorter or the same bond lengths?
BeO, BeS, BN, BP, AlN, AlP.

11.2 Stishovite is a high-pressure modification of SiO$_2$ having the rutile structure. Should it have longer or shorter Si–O bond lengths than quartz?

11.3 Whereas AgCl has the NaCl structure, AgI has the sphalerite structure. Could you imagine conditions under which both compounds would have the same structure?

11.4 What is the coordination number of the iodine atoms in β-Cu$_2$HgI$_4$?

11.5 If well-crystallized Hg$_2$C could be made, what structure should it have?

12   Polyanionic and Polycationic Compounds. Zintl Phases

The compounds dealt with first in this chapter belong to the *normal valence compounds*; these are compounds that fulfil the classic valence concept of stable eight-electron shells. They include not only the numerous molecular compounds of nonmetals, but also compounds made up of elements from the left side with elements from the right side of the ZINTL line. The ZINTL line is a delimiting line that runs in the periodic table of the elements between the third and the fourth main group. According to the classical concepts, such compounds consist of ions, for exmaple NaCl, K_2S, Mg_2Sn, Ba_3Bi_2. Judging by the composition, however, in many cases the octet rule seems to be violated as, for example, in $CaSi_2$ or NaP. This impression is erroneous: the octet rule is still being fulfilled; the formation of covalent bonds renders it possible. In $CaSi_2$ the Si atoms are joined in layers as in gray arsenic (Si^- and As are isoelectronic), and in NaP the phosphorus atoms form helical chains analogous to polymer sulfur (P^- and S are isoelectronic). Whether a compound fulfils the octet rule can only be decided when its structure is known.

12.1   The Generalized $8 - N$ Rule

The octet principle can be expressed as a formula by the *generalized $8-N$ rule* (E. MOOSER & W.B. PEARSON [99]). We restrict our considerations to binary compounds, and presuppose the following:

1. Let X be an element of the fourth to seventh main group of the periodic table, i.e. an element that tends to attain the electronic configuration of the following noble gas by taking up electrons (the heavy elements of the third main group may also be included). An X atom has $e(X)$ valence electrons.

2. The electrons needed to fill up the electron octet at X are supplied by the more electropositive element M. An M atom has $e(M)$ valence electrons.

The composition being M_mX_x, $8x$ electrons are required in order to achieve the octet shells for the x X atoms:

$$m \cdot e(M) + x \cdot e(X) = 8x \tag{13}$$

If covalent bonds exist between M atoms, then not all of the $e(M)$ electrons of M can be turned over to X, and the number $e(M)$ in equation (13) must be reduced by the number $b(MM)$ of covalent bonds per M atom; if the M atoms retain nonbonding electrons (lone electron pairs as for Tl^+), then $e(M)$ must also be reduced by the number E of these electrons. On the other hand, the X atoms require fewer electrons if they take part in covalent bonds with each other; the number $e(X)$ can be increased by the number $b(XX)$ of covalent bonds per X atom:

$$m[e(M) - b(MM) - E] + x[e(X) + b(XX)] = 8x \tag{14}$$

By rearrangement of this equation we obtain:

$$\frac{m \cdot e(M) + x \cdot e(X)}{x} = 8 + \frac{m[b(MM) + E] - x \cdot b(XX)}{x} \tag{15}$$

We define the *valence electron concentration per anion*, VEC(X), as the total number of *all* valence electrons in relation to the number of anionic atoms:

$$VEC(X) = \frac{m \cdot e(M) + x \cdot e(X)}{x} \tag{16}$$

By substituting equation (15) into equation (16) and solving for $b(XX)$ we obtain:

$$b(XX) = 8 - VEC(X) + \frac{m}{x}[b(MM) + E] \tag{17}$$

Equation (17) represents the generalized $8 - N$ rule. Compared to the simple $8 - N$ rule (p. 49), it is enlarged by the term $\frac{m}{x}[b(MM) + E]$, and VEC(X) has taken the place of the main group number N. The following specialized cases are of importance:

1. Elements. For pure elements that belong to the right side of the ZINTL line, we have $m = 0$, VEC(X)$= e(X) = N$, and equation (17) becomes:

$$b(XX) = 8 - VEC(X) = 8 - N \tag{18}$$

This is none other than the simple $8 - N$ rule. For example, in sulfur ($N = 6$) the number of covalent bonds per S atom is $b(SS) = 8 - N = 2$.

2. Polyanionic compounds. Frequently, the M atoms lose all their valence electrons to the X atoms, i.e. no cation–cation bonds occur and no nonbonding electrons remain at the cations, $b(MM)=0$ and $E = 0$. Equation (17) then becomes:

$$b(XX) = 8 - VEC(X) \tag{19}$$

This once again is the $8 - N$ rule, but only for the anionic component of the compound. For example: Na_2O_2; VEC(O) $= 7$; $b(OO) = 8 - 7 = 1$, there is one covalent bond per O atom.

By comparing equations (18) and (19) we can deduce:

The geometric arrangement of the atoms in a polyanionic compound corresponds to the arrangement in the structures of the elements of the fourth to seventh main group when the number of covalent bonds per atom $b(XX)$ is equal. According to this conception, put forward by E. ZINTL and further developed by W. KLEMM and E. BUSMANN, the more electronegative partner in a compound is treated like that element which disposes of the same number of electrons. This statement is therefore a specialized case of the general rule according to which isoelectronic atom groups adopt the same kind of structures.

3. Polycationic compounds. Provided that no covalent bonds occur between the anionic atoms, $b(XX)=0$, equation (17) becomes:

$$b(MM) + E = \frac{x}{m}[VEC(X) - 8] \qquad (20)$$

When applying this equation, note that for the calculation of VEC(X) according to equation (16) *all* valence electrons have to be considered, including those that take part in M–M bonds.

For example: Hg_2Cl_2; $e(Hg) = 2$; $VEC(Cl) = 9$; $b(HgHg) = 1$ (when the 10 d electrons of an Hg atom are also considered as being valence electrons, then $VEC(Cl) = 19$, $E = 10$, $b(HgHg) = 1$).

4. Simple ionic compounds, i.e. compounds having no covalent bonds, $b(MM)= b(XX) = E = 0$. Equation (17) becomes:

$$VEC(X) = 8$$

which is the octet rule.

We now can classify compounds according to the value of VEC(X). Since $b(MM)$, E and $b(XX)$ cannot adopt negative values, VEC(X) in equation (19) must be smaller than 8, and in equation (20) it must be greater than 8. We thus deduce the criterion:

$$VEC(X) < 8 \qquad \text{polyanionic}$$
$$VEC(X) = 8 \qquad \text{simple ionic}$$
$$VEC(X) > 8 \qquad \text{polycationic}$$

As VEC(X) is easy to calculate according to equation (16), we can quickly estimate the kind of bonding in a compound, for example:

polycationic	VEC(X)	polyanionic	VEC(X)	simple ionic	VEC(X)
Ti_2S	14	Ca_5Si_3	$7\frac{1}{3}$	Mg_2Sn	8
$MoCl_2$	10	Sr_2Sb_3	$6\frac{1}{3}$	Na_3P	8
$Cs_{11}O_3$	$9\frac{2}{3}$	CaSi	6	wrong:	
GaSe	9	KGe	5	InBi	8

As we can see from the last entry in this table, we have deduced only a *rule*. In InBi there are Bi–Bi contacts and it has metallic properties. Further examples that do not fulfil the rule are LiPb (Pb atoms surrounded only by Li) and K_8Ge_{46}. In the latter, all Ge atoms have four covalent bonds; they form a wide-meshed framework that encloses the K^+ ions (Fig. 112, p. 182); the electrons donated by the potassium atoms are not taken over by the germanium, and instead they form a band. In a way, this is kind of a solid solution, with germanium as "solvent" for K^+ and "solvated" electrons. K_8Ge_{46} has metallic properties. In the sense of the $8 - N$ rule the metallic electrons can be "captured": in $K_8Ga_8Ge_{38}$, which has the same structure, all the electrons of the potassium are required for the framework, and it is a semiconductor. In spite of the exceptions, the conception has turned out to be very fruitful, especially in the context of understanding the ZINTL phases.

12.2 Polyanionic Compounds, Zintl Phases

Table 16 lists some binary polyanionic compounds, arranged according to the valence electron concentration per anion atom. Only compounds with integral values for VEC(X) are listed. In agreement with the above-mentioned rule, in fact structures like those of pure elements with the corresponding numbers of valence electrons occur for the anionic components. However, the variety of structures is considerably larger than for the pure elements. For example, three-bonded atoms not only occur in the layer structures as in phosphorus and arsenic, but also in several other connection patterns (Fig. 63). This seems reasonable, since the anionic grid has to make allowance for the space requirements of the cations. $CaSi_2$, for instance, has layers $(Si^-)_\infty$ as in arsenic; $SrSi_2$, however, has a complicated network structure with three-bonded Si atoms. Under pressure, both $CaSi_2$ and $SrSi_2$ are transformed to the α-$ThSi_2$ type, with yet another kind of network of three-bonded Si atoms. Contrary to expectations based on the $8-N$ rule, the Si atoms in the α-$ThSi_2$ type do not have pyramidal coordination, but planar coordination (in $SrSi_2$ coordination is nearly planar).

The calculation of VEC(X) for many compounds results in non-integral numbers. According to equation (19) fractional numbers then also result for the number $b(XX)$ of covalent bonds. This happens when structurally different atoms occur in the anion. The following examples help to illustrate this:

Na_2S_3: with VEC(X)$=\frac{20}{3}$ we obtain $b(XX)=\frac{4}{3}$. This is due to the chain struc-

ture of the S_3^{2-} ion. For the two terminal atoms we have $b(XX)=1$, and for the central one $b(XX)=2$; the average is $(2\cdot 1+2)/3=\frac{4}{3}$. For unbranched chains with specific lengths as in polysulfides S_n^{2-}, $6<$VEC(X)<7 holds as long as no

Table 16 Examples of polyanionic compounds which have integral valence electron concentrations per anion atom

Example	VEC(X)	$b(XX)$	structure of the anion part
Li_2S_2	7	1	S_2^{2-} pairs as in Cl_2
FeS_2 $\}$ $FeAsS$	7	1	S_2^{2-} and AsS^{3-} pairs
NiP	7	1	P_2^{4-} pairs
CaSi	6	2	zigzag chains
LiAs	6	2	helical chains
$CoAs_3$	6	2	four-membered As_4^{4-} rings
InP_3	6	2	P_6^{6-} rings (chair) as in S_6
$CaSi_2$	5	3	undulated layers as in α-As
HP-$SrSi_2$	5	3	interconnected zigzag chains
K_4Ge_4	5	3	Ge_4^{4-} tetrahedra as in P_4
CaC_2	5	3	C_2^{2-} pairs as in N_2
NaTl	4	4	diamond-like
$SrGa_2$	4	4	graphite-like

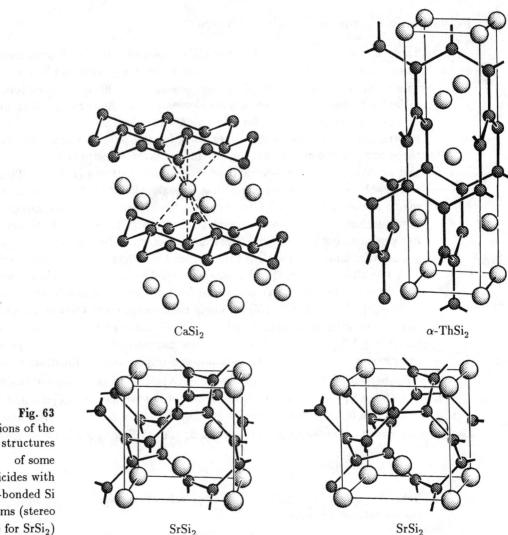

CaSi$_2$ α-ThSi$_2$

Fig. 63
Sections of the
structures
of some
polysilicides with
three-bonded Si
atoms (stereo
image for SrSi$_2$)

SrSi$_2$ SrSi$_2$

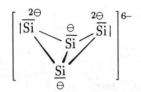

multiple bonds occur. When there are multiple bonds, VEC(X)$<$ 6 is possible, e.g. VEC(N) = 5.33 for the azide ion, $\langle$N=N=N$\rangle^-$.
Ba$_3$Si$_4$: VEC(X) = $\frac{11}{2}$, b(XX) = $\frac{5}{2}$. An average value of $2\frac{1}{2}$ covalent bonds per Si atom results when half of the Si atoms are bonded with two covalent bonds, and the other half with three covalent bonds. This corresponds to the real structure.

The number of negative charges of the anion can also be counted in the following way: every atom of the N-th main group that participates in exactly $8-N$ covalent bonds obtains a formal charge of zero; for every bond less than $8-N$ it obtains a negative formal charge. A four-bonded silicon atom thus obtains a formal charge of 0, a three-bonded one obtains 1$\ominus$ and a two-bonded one obtains 2$\ominus$. The sum of all formal charges is equal to the ionic charge.

Sometimes rather complicated structures occur in the anionic part of a structure. For instance, approximately 50 different binary polyphosphides are known only for the alkali and alkaline earth metals which, in part, also adopt different modifications; in addition, there are more than 120 binary polyphosphides of other metals [101]. Fig. 64 conveys an impression of how manifold the structures are. Apart from simple chains and rings, cages like those in sulfides such as As_4S_4 and P_4S_3 (and others) have been observed (every P atom that substitutes an S atom is to be taken as a $P^{\ominus}$). Layer structures can be regarded as sections of the structure of black phosphorus; other structures correspond to fragments of the structure of HITTORF's phosphorus. The diversity in polyarsenides, polyantimonides and polysilicides is just as complicated.

Binary polyanionic compounds can frequently be synthesized directly from the elements. In some cases, intact cage-like anions can be extracted from the solids when a complexing ligand is offered for the cation. For example, the Na^+ ions of Na_2Sn_5 can be captured by cryptand molecules, $\rightarrow [NaCrypt^+]_2Sn_5^{2-}$. Cryptands like $N(C_2H_4OC_2H_4OC_2H_4)_3N$ enclose the alkali ion. For some of the cage-like anions the kind of bonding is consistent with the preceding statements, but for some others they do not seem to apply. The ionic charges of P_7^{3-} or As_{11}^{3-} correspond exactly to the numbers of two-bonded $P^{\ominus}$ or $As^{\ominus}$ atoms (Fig. 64). But for Sn_5^{2-} or Sn_9^{4-} this is not so clear. For Sn_5^{2-} several models have been put forward; the 22 valence electrons can be accommodated exactly when multicenter bonds as in boranes are assumed, i.e. 6 ($= n + 1$, cf. p. 133) bonds in the cluster and one lone electron pair at each Sn atom; they can also be accommodated when, in addition to the lone electron pairs, a 2e3c bond is assumed for every one of the six faces of the trigonal bipyramid. However, the assumption of multicenter bonds, which are normally indispensable only for electron deficient compounds, appears to be inconsistent for a species that has no serious electron deficiency (there are lone electron pairs!). If one shakes off the prejudice that all pyramidally coordinated Sn atoms must have a lone electron pair pointing outwards, then Sn_5^{2-} and Sn_9^{4-} fit well into the simple valence rules: a four-bonded Sn atom is uncharged, a three-bonded one with a lone electron pair has a negative formal charge, and a five-bonded one without a lone electron pair also has a negative formal charge (the five-bonded one violating the octet rule):

Sn_5^{2-}

trigonal bipyramid

Sn_9^{4-}

capped square antiprism

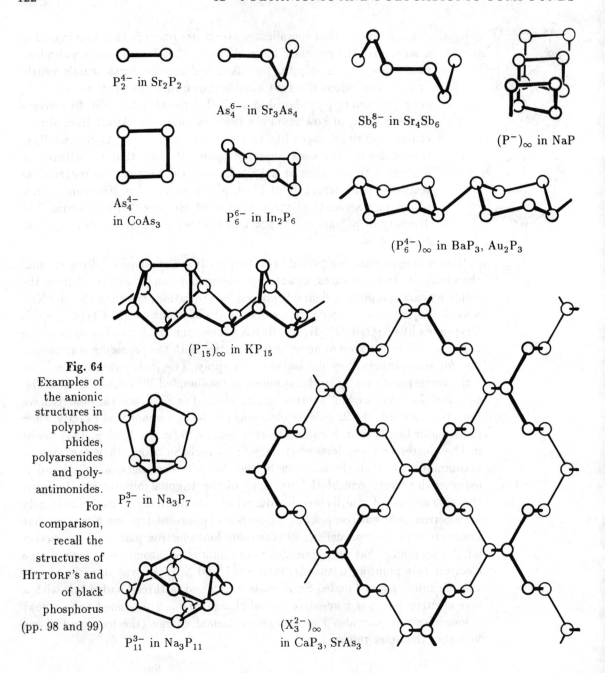

Fig. 64
Examples of the anionic structures in polyphosphides, polyarsenides and polyantimonides. For comparison, recall the structures of HITTORF's and of black phosphorus (pp. 98 and 99)

P_2^{4-} in Sr_2P_2

As_4^{6-} in Sr_3As_4

Sb_6^{8-} in Sr_4Sb_6

$(P^-)_\infty$ in NaP

As_4^{4-} in $CoAs_3$

P_6^{6-} in In_2P_6

$(P_6^{4-})_\infty$ in BaP_3, Au_2P_3

$(P_{15}^-)_\infty$ in KP_{15}

P_7^{3-} in Na_3P_7

P_{11}^{3-} in Na_3P_{11}

$(X_3^{2-})_\infty$ in CaP_3, $SrAs_3$

The relation between structure, stability, bonding and the number of electrons is by no means clear for polyanions of this kind [102]. The ion Sn_9^{3-} has the structure of a tricapped trigonal prism, the same as the ion Bi_9^{5+} that is isoelectronic to Sn_9^{4-} [111]. A Ge_9^{2-} ion with a similar structure is also known.

Zintl Phases

Many of the compounds presented in the preceding paragraphs belong to the **Zintl phases**. This is a class of compounds consisting of an electropositive, cationic component (alkali metal, alkaline earth metal, lanthanoid) and an anionic component of main group elements of moderate electronegativity. The anionic part of the structure fulfils the simple concept of normal valence compounds. Nevertheless, the compounds are not salt-like, but have metallic properties, especially metallic luster. However, they are not "full-value" metals; instead of being metallic-ductile, many of them are brittle. As far as the electrical properties have been studied, mostly semiconductivity has been found. There are many analogies with the half-metallic elements: in the structures of germanium, α-tin, arsenic, antimony, bismuth, selenium and tellurium the $8 - N$ rule can be discerned; although these elements can be considered to be normal valence compounds, they show metallic luster, but they are brittle and are semiconductors or moderate metallic conductors.

The classic example of a ZINTL phase is the compound NaTl which can be interpreted as Na^+Tl^-; its thallium partial structure has the diamond structure (Fig. 65). In NaTl the Tl–Tl bonds are significantly shorter than the contact distances in metallic thallium (324 instead of 343 pm, albeit with a reduced coordination number). Although the valence electron concentration is the same, the Ga^- particles in $SrGa_2$ do not form a diamond-like structure, but layers as in graphite (AlB_2 type; AlB_2 itself does not fulfil the octet rule). All compounds listed in Table 16 with the exception of Li_2S_2 and CaC_2 are ZINTL phases (recall the golden luster of pyrite, FeS_2). The number of known ZINTL phases is enormous; many of them have been studied by HERBERT SCHÄFER, H.G.V. SCHNERING and J.D. CORBETT [101,102,106,107,108,111].

In ternary ZINTL phases the anionic part of the structures resembles halo or oxo anions or molecular halides [109]. For example, in Ba_4SiAs_4 there are tetrahedral $SiAs_4^{8-}$ particles that are isostructural to $SiBr_4$ molecules. In

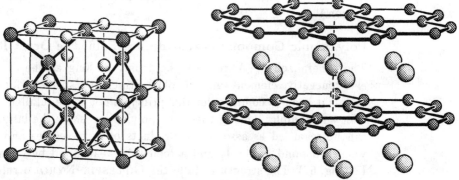

Fig. 65
Left: unit cell of NaTl. The plotted bonds of the thallium partial structure correspond to the C–C bonds in diamond. Right: section of the structure of $SrGa_2$ (AlB_2 type)

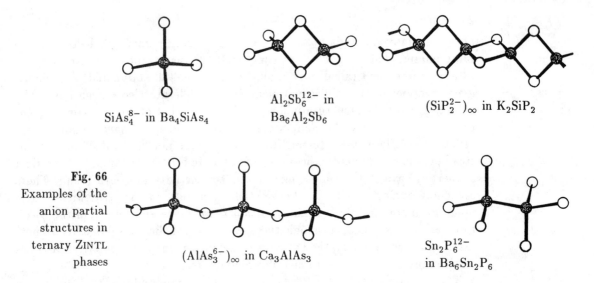

$SiAs_4^{8-}$ in Ba_4SiAs_4

$Al_2Sb_6^{12-}$ in
$Ba_6Al_2Sb_6$

$(SiP_2^{2-})_\infty$ in K_2SiP_2

Fig. 66
Examples of the
anion partial
structures in
ternary ZINTL
phases

$(AlAs_3^{6-})_\infty$ in Ca_3AlAs_3

$Sn_2P_6^{12-}$
in $Ba_6Sn_2P_6$

Ba$_3$AlSb$_3$ dimer groups $Al_2Sb_6^{12-}$ are present, with a structure as in Al_2Cl_6 molecules (Fig. 66). Ca_3AlAs_3 contains polymer chains of linked tetrahedra $(AlAs_3^{6-})_\infty$ as in chain silicates $(SiO_3^{2-})_\infty$. The compound $Ca_{14}AlSb_{11}$ = $[Ca^{2+}]_{14}[Sb^{3-}]_4[Sb_3^{7-}][AlSb_4^{9-}]$ contains three kinds of anions, namely single ions Sb^{3-}, ions Sb_3^{7-} that are isostructural to I_3^-, and tetrahedral $AlSb_4^{9-}$ ions. $Ba_6Sn_2P_6$ has $Sn_2P_6^{12-}$ particles with an Sn–Sn bond; their structure is like that of ethane. Also, complicated chains and frameworks are known that are reminiscent of the manifold structures of the silicates; however, the possible varieties are by far greater than for silicates because the anionic component is not restricted to the linking of SiO_4 tetrahedra.

The octet principle, primitive as it may appear, has not only been applied very succesfully to the half-metallic ZINTL phases, but it is also theoretically well-founded (requiring a lot of computational expenditure). Evading the purely metallic state with delocalized electrons in favor of electrons more localized in the anionic partial structure can be understood as PEIERLS distortion (cf. section 9.4).

Polyanionic Compounds that do not Fulfil the Octet Rule

The generalized $8-N$ rule can hold only as long as the atoms of the more electronegative element fulfil the octet principle. Especially for the heavier non-metals it is quite common for this principle not to be fulfilled. The polyhalides offer an example. Among these the polyiodides show the largest variety. They can be regarded as association products of I_2 molecules and I^- ions, with a weakened bond in the I_2 and a relatively weak bond between the I_2 and the I^- (Fig. 67). The structures fulfil the GILLESPIE–NYHOLM rules. This also applies to the polytellurides, among which associations to larger units have been observed [105]; the Te_5^{6-} ion, for example, is square like the BrF_4^- ion (Fig. 68).

Fig. 67
Structures
of some
polyiodides. The
I_2 building units
are in bold face.
Distances in pm.
For comparison:
molecule I—I 268
pm, VAN DER
WAALS distance
I···I 396 pm

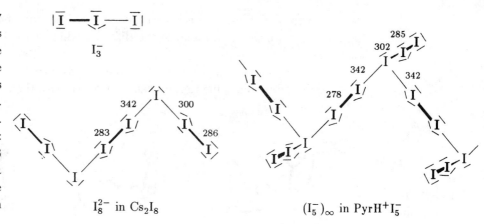

I_3^-

I_8^{2-} in Cs_2I_8

$(I_5^-)_\infty$ in $PyrH^+I_5^-$

Fig. 68
Structures
of some
polytellurides.
Lone electron
pairs are marked
as double dots.
Te_5^{2-} is also
known in the
form of simple
chain structures
as in S_5^{2-}

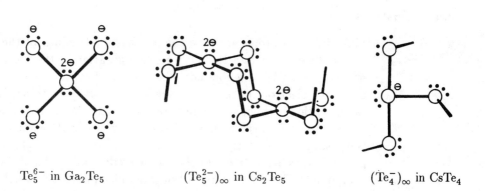

Te_5^{6-} in Ga_2Te_5

$(Te_5^{2-})_\infty$ in Cs_2Te_5

$(Te_4^-)_\infty$ in $CsTe_4$

12.3   Polycationic Compounds

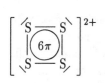

The number of known polycationic compounds of main group elements is far less than that of polyanionic compounds. Examples include the chalcogen cations S_4^{2+}, S_8^{2+}, Se_{10}^{2+} and Te_6^{4+} that are obtained when the elements react with Lewis acids under oxidizing conditions [100]. The ions S_4^{2+}, Se_4^{2+} and Te_4^{2+} have a square structure that can be assumed to have a 6π electron system.

The structures of S_8^{2+} and Se_8^{2+} can be interpreted with the $8-N$ rule: a bond is generated across an S_8 ring, resulting in two atoms having three bonds and one positive formal charge each (Fig. 69). The new bond, indeed, is remarkably long (283 pm as compared to 205 pm for the other bonds), but the occurrence of abnormally long S–S bonds is also known for some other sulfur compounds. The structure of the $Te_3S_3^{2+}$ ion can also be understood in terms of the $8-N$ rule, whereas the trigonal-prismatic structure of the Te_6^{4+} ion does not comply with it (its structure would suit a Te_6^{6+} ion). The rather different bond lengths of the Te_6^{4+} ion are remarkable (Fig. 69); since there should be one bond less, it is as if the ion could not "make up its mind" which of the three long bonds should be cleaved, i.e. resonance formulas with two bonds distributed to three prism edges have to be drawn. There exist other polycations with bonding relations that are not quite clear, for instance the Bi_9^{5+} ion which is isoelectronic to Sn_9^{4-} (cf. p. 121).

In a broader sense numerous cluster compounds can be considered to be polycationic compounds; due to their variety we deal with them next in a section of their own.

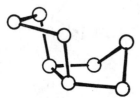
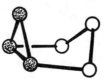
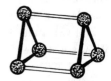

Fig. 69
Structures of the
ions S_8^{2+}, $Te_3S_3^{2+}$
and Te_6^{4+}

12.4   Cluster Compounds

Links between atoms serve to compensate for the lack of the electrons which are necessary to attain the electron configuration of the next noble gas in the periodic table. With a common electron pair between two atoms each of them gains one electron in its valence shell. As the two electrons link two "centers" (atoms), this is called a two-electron two-center bond or, for short, *2e2c* bond. If, for an element, the number of available partner atoms of a different element is not sufficient to fill the valence shell, atoms of the same element combine with each other, as is the case for polyanionic compounds or for the numerous organic compounds. For the majority of polyanionic compounds a sufficient number of electrons is available to satisfy the demand for electrons with the aid of *2e2c* bonds. Therefore, the generalized $8 - N$ rule usually is fulfilled for polyanionic compounds.

For more electropositive elements, which have an inferior number of valence electrons in the first place, and which in addition have to supply electrons to a more electronegative partner, the number of available electrons is rather small. They can gain electrons in two ways: first, as far as possible, by complexation, i.e. by the acquisition of ligands; and second, by combining their own atoms with each other. This can result in the formation of clusters. A cluster is an accumulation of three or more atoms of the same element or of similar elements that are directly linked with each other. If the accumulation of atoms yields a sufficient number of electrons to allow for one electron pair for every connecting line between two adjacent atoms, then each of these lines can be taken to be a *2e2c* bond just as in a common valence bond formula. Clusters of this kind have been called *electron precise*.

For low values of the valence electron concentration (VEC< 4 for main group elements), covalent *2e2c* bonds are not sufficient to overcome the electron deficiency. We have the case of "electron deficient compounds". For these, relief comes from *multicenter bonds*. In a two-electron three-center bond (*2e3c*) three atoms share an electron pair. An even larger number of atoms can share one electron pair. With increasing numbers of atoms sharing the same electron pair, each atom is less tightly bonded. The electron pair in a 2e3c bond essentially is located in the center of the triangle defined by the three atoms:

The location of electrons linking more than three centers cannot be illustrated as easily. The simple, descriptive models must give way to the theoretical treatment by molecular orbital theory. With its aid, however, certain electron counting rules have been deduced for cluster compounds that set up relations between the structure and the number of valence electrons [94,95,96,97,98].

Completely closed, convex, single-shell clusters are called *closo* clusters; their atoms form a polyhedron. If the polyhedron has only triangular faces, it is also called a deltahedron. Depending on the number of available electrons, we can distinguish four general bonding types for *closo* clusters:

1. Electron precise clusters with exactly one electron pair per polyhedron edge;

2. Clusters with one 2e3c bond for every triangular face;

3. Clusters for which the WADE rules discussed on p. 133 apply;

4. Clusters not matching any of these patterns.

Electron Precise Clusters

Molecules such as P_4 and the polyanionic clusters such as Si_4^{4-} or As_7^{3-} that are discussed in section 12.2 are representatives of electron precise closo clusters. Organic cage molecules like tetrahedrane (C_4R_4), prismane (C_6H_6), cubane (C_8H_8), and dodecahedrane ($C_{20}H_{20}$) also belong to this kind of clusters.

Numerous clusters with electron numbers that account for exactly one electron pair per polyhedron edge are also known for the more electron rich transition group elements (beginning with group six). In addition, every cluster atom obtains electrons from coordinated ligands, with a tendency to attain a total of 18 valence electrons per atom. The easiest way to count the number of electrons is to start from uncharged metal atoms and uncharged ligands. Ligands such as NH_3, PR_3, and CO supply two electrons. Nonbridging halogen atoms, H atoms and groups such as SiR_3 supply one electron (for halogen atoms this amounts to the same as assuming a Hal^- ligand that makes available two electrons, but that had previously obtained an electron from a metal atom). A μ_2-bridging halogen atom supplies three electrons (one as before plus one of its lone electron pairs), and a μ_3-bridging halogen atom five. Table 17 lists how many electrons are to be taken into account for some ligands.

Table 17 Number of electrons supplied by ligands to metal atoms in complexes when the metal atoms are considered to be uncharged.
μ_1 = terminal ligand, μ_2 = ligand bridging two atoms, μ_3 = ligand bridging three atoms; *int* = interstitial atom inside a cluster

ligand		electrons	ligand		electrons
H	μ_1	1	NR_3	μ_1	2
H	μ_2	1	NCR	μ_1	2
H	μ_3	1	NO	μ_1	3
CO	μ_1	2	PR_3	μ_1	2
CO	μ_2	2	OR	μ_1	1
CS	μ_1	2	OR	μ_2	3
CR_2	μ_1	2	OR_2	μ_1	2
$\eta^2\text{-}C_2R_4$	μ_1	2	O, S, Se, Te	μ_1	0
$\eta^2\text{-}C_2R_2$	μ_1	2	O, S, Se, Te	μ_2	2
$\eta^5\text{-}C_5R_5$	μ_1	5	O, S, Se, Te	μ_3	4
$\eta^6\text{-}C_6R_6$	μ_1	6	O, S	*int*	6
C	*int*	4	F, Cl, Br, I	μ_1	1
SiR_3	μ_1	2	F, Cl, Br, I	μ_2	3
N, P	*int*	5	Cl, Br, I	μ_3	5

The electrons supplied by the ligands and the valence electrons of the n metal atoms of an M_n cluster are added to a total electron number g. The number of M–M bonds (polyedron edges) then is:

$$\text{main group element clusters:} \quad b \;=\; \frac{1}{2}(8n - g) \tag{21}$$

$$\text{transition element clusters:} \quad b \;=\; \frac{1}{2}(18n - g) \tag{22}$$

This mode of calculation has been called the "EAN rule" (effective atomic number rule [97]). It is valid for arbitrary metal clusters (*closo* and others) if the number of electrons is sufficient to assign one electron pair for every M–M connecting line between adjacent atoms, and if the octet rule or the 18-electron rule is fulfilled for main group elements or for transition group elements, respectively. The number of bonds b calculated this way is a limiting value: the number of polyhedron edges in the cluster can be greater than or equal to b, but never smaller. If it is equal, the cluster is electron precise.

Since an M atom gains one electron per M–M bond, the calculation can also be performed in the following way: the total number g of valence electrons of the cluster must be equal to:

$$\text{main group element clusters:} \quad g \;=\; 7n_1 + 6n_2 + 5n_3 + 4n_4 \tag{23}$$

$$\text{transition element clusters:} \quad g \;=\; 17n_1 + 16n_2 + 15n_3 + 14n_4 \tag{24}$$

n_1, n_2, n_3, and n_4 are the numbers of polyhedron vertices at which 1, 2, 3 or 4 polyhedron edges (M–M bonds) meet, respectively. Polyhedra with five or more edges per vertex are generally not electron precise (for this reason no numbers n_5, n_6,... occur in the equations). Therefore, the expected valence electron numbers for some simple polyhedra are:

	main group elements	transition group elements
triangle	18	48
tetrahedron	20	60
trigonal bipyramid	22	72
octahedron	24	84
trigonal prism	30	90
cube	40	120

As an exercise, one could calculate the numbers for some of the polyanionic compounds in section 12.2. Further examples include:

$$(CO)_4$$
$$Os$$
$$(OC)_4Os\!-\!Os(CO)_4$$

$$Os_3(CO)_{12} \quad \begin{array}{lr} 3\ Os & 3 \times 8 = 24 \\ 12\ CO & 12 \times 2 = 24 \\ \hline g = & 48 \end{array} = 16n_2$$

$$b = \tfrac{1}{2}(18 \times 3 - 48) = 3$$

$$(CO)_3$$
$$Ir$$
$$(OC)_3Ir\!-\!\!\!\mid\!\!\!-Ir(CO)_3$$
$$Ir$$
$$(CO)_3$$

$$Ir_4(CO)_{12} \quad \begin{array}{lr} 4\ Ir & 4 \times 9 = 36 \\ 12\ CO & 12 \times 2 = 24 \\ \hline g = & 60 \end{array} = 15n_3$$

$$b = \tfrac{1}{2}(18 \times 4 - 60) = 6$$

$$(OC)_3Os\!\!-\!\!\begin{array}{c} Os(CO)_3 \\ Os(CO)_3 \\ P \\ Os(CO)_3 \\ Os(CO)_3 \end{array}$$
$$(OC)_3Os$$

$$[Os_6(CO)_{18}P]^- \quad \begin{array}{lr} 6\ Os & 6 \times 8 = 48 \\ 18\ CO & 18 \times 2 = 36 \\ P & 5 \\ \text{charge} & 1 \\ \hline g = & 90 \end{array} = 15n_3$$

$$b = \tfrac{1}{2}(18 \times 6 - 90) = 9$$

$$[Mo_6Cl_{14}]^{2-} \quad \begin{array}{lr} 6\ Mo & 6 \times 6 = 36 \\ 8\ \mu_3\text{-Cl} & 8 \times 5 = 40 \\ 6\ \mu_1\text{-Cl} & 6 \times 1 = 6 \\ \text{charge} & 2 \\ \hline g = & 84 \end{array} = 14n_4$$

$$b = \tfrac{1}{2}(18 \times 6 - 84) = 12$$

The cluster mentioned last, $[Mo_6Cl_{14}]^{2-}$, also occurs in $MoCl_2$. It consists of an Mo_6 octahedron inscribed in a Cl_8 cube; every one of the eight Cl atoms of the cube is situated on top of one of the octahedron faces and is coordinated to

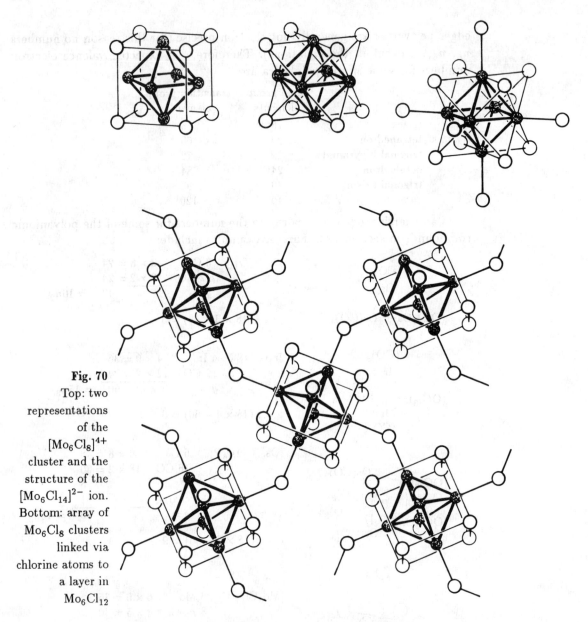

Fig. 70
Top: two representations of the $[Mo_6Cl_8]^{4+}$ cluster and the structure of the $[Mo_6Cl_{14}]^{2-}$ ion. Bottom: array of Mo_6Cl_8 clusters linked via chlorine atoms to a layer in Mo_6Cl_{12}

three molybdenum atoms (Fig. 70). The formula $[Mo_6Cl_8]^{4+}$ applies to this unit; in it, every Mo atom is still short of two electrons it needs to attain 18 valence electrons. They are supplied by the six Cl^- ions bonded at each octahedron vertex. This also applies to $MoCl_2$, but there are only four Cl^- per cluster; however, two of them act as bridging ligands between clusters, corresponding to the formula $[Mo_6Cl_8]Cl_{2/1}Cl_{4/2}$ (Fig. 70).

The situation is very similar in the CHEVREL phases [110]. These are ternary molybdenum chalcogenides $A_x[Mo_6X_8]$ (A = metal, X = S, Se) that have attracted much attention because of their physical properties, especially as su-

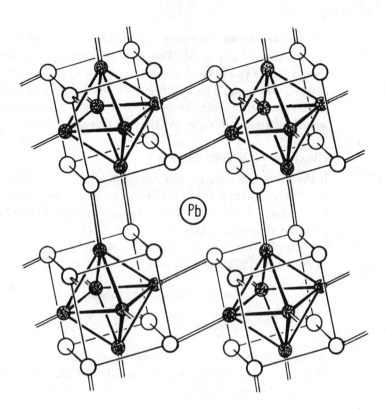

Fig. 71
Association of
Mo_6S_8 clusters in
the CHEVREL
phase $PbMo_6S_8$

perconductors. The "parent compound" is $PbMo_6S_8$; it contains Mo_6S_8 clusters that are linked with each other in such a way that the free coordination sites of one cluster are occupied by sulfur atoms of adjacent clusters (Fig. 71). The electric properties of CHEVREL phases depend on the number of valence electrons. With 24 electrons per cluster (one electron pair for every edge of the Mo_6 octahedron) the cluster is electron-precise, the valence band is fully occupied and the compounds are semiconductors, as, for example, $(Mo_4Ru_2)Se_8$ (it has two Mo atoms substituted by Ru atoms in the cluster). In $PbMo_6S_8$ there are only 22 electrons per cluster; the "electron holes" facilitate a better electrical conductivity; below 14 K it becomes a superconductor. By incorporating other elements in the cluster and by the choice of the electron donating element A, the number of electrons in the cluster can be varied within certain limits (20 to 24 electrons for the octahedral skeleton). With the lower electron numbers the weakened cluster bonds show up in trigonally elongated octahedra.

If electrons are added to an electron precise cluster, cleavage of bonds is to be expected according to equation (21) or (22); for every additional electron pair g increases by 2 and b decreases by 1. The Si_4^{6-} ion presented on p. 120 is an example; it can be thought of having been formed from a tetrahedral Si_4^{4-} by the addition of two electrons. Another example is $Os_3(CO)_{12}(SiCl_3)_2$ with a linear Os–Os–Os group; by attaching two $SiCl_3$ groups to triangular $Os_3(CO)_{12}$, two more electrons are supplied, and one Os–Os bond has to be cleaved.

However, certain polyhedra allow the inclusion of another electron pair without cleavage of any bond. This applies especially for octahedral clusters which should have 84 valence electrons according to equation (24), but they frequently have 86 electrons. The additional electron pair assumes a bonding action as a six-center bond inside the octahedron [100]. An octahedral cluster with 86 valence electrons fulfils the WADE rule discussed below.

Clusters with 2e3c Bonds

If there are not enough electrons for all of the polyhedron edges, 2e3c bonds on the triangular polyhedron faces can be the next best solution to compensate for the lack of electrons. This solution is only possible for deltahedra that have no more than four edges (and faces) meeting at any vertex. These include especially the tetrahedron, trigonal bipyramid and octahedron. Relations similar to equations (21) and (22) can be specified:

$$\text{main group element clusters:} \quad d = \frac{1}{4}(8n - g) \tag{25}$$

$$\text{transition element clusters:} \quad d = \frac{1}{4}(18n - g) \tag{26}$$

Again, n is the number of cluster atoms and g is the total number of valence electrons of the cluster. d is the number of triangular polyhedron faces having 2e3c bonds. The equation is valid only when 2e3c bonds are the only kind of bonds to be assumed in the cluster.

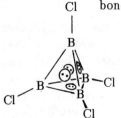

For example, the bonding in B_4Cl_4 can be interpreted in the following way: every boron atom takes part in four bonds, one 2e2c B–Cl bond and three 2e3c bonds on the faces of the B_4 tetrahedron. This way every boron atom attains an electron octet. The number of valence electrons is $g = 4 \cdot 3 + 4 = 16$, and the number of 2e3c bonds is calculated to be $d = \frac{1}{4} \cdot 16 = 4$.

In the $Nb_6Cl_{18}^{4-}$ ion the octahedral Nb_6 cluster can be assumed to have eight 2e3c bonds on its eight octahedron faces. A chlorine atom bonded with two Nb atoms is situated next to each octahedron edge. This makes twelve Cl atoms in a $Nb_6Cl_{12}^{2+}$ unit. The remaining six Cl^- ions are terminally bonded to the octahedron vertices (Fig. 72). The number of valence electrons is:

6 Nb	$6 \times 5 =$	30
12 μ_2-Cl	$12 \times 3 =$	36
6 μ_1-Cl	$6 \times 1 =$	6
charge		4
		76

$$d = \tfrac{1}{4}(18 \times 6 - 76) = 8$$

For each Nb atom the situation is the same as in the $Mo_6Cl_{14}^{2-}$ ion: the metal atom is surrounded by five Cl atoms and is involved in four metal–metal bonds in the cluster. However, the MCl_5 unit is rotated with respect to the octahedron: Cl atoms on top of the Mo_6 octahedron faces become Cl atoms on top of the Nb_6 octahedron edges, and the bonding electron pairs switch over from the

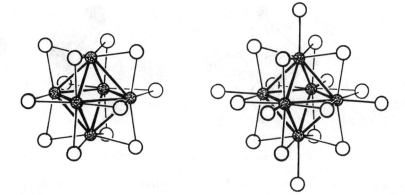

Fig. 72
Structures of the
$Nb_6Cl_{12}^{2+}$ cluster
and the $Nb_6Cl_{18}^{4-}$
ion

edges to the faces. In both cases the valence electrons for a metal atom add up to 18. In Nb_6Cl_{14} the $Nb_6Cl_{12}^{2+}$ clusters are associated via intervening chlorine atoms, similar to Mo_6Cl_{12}.

Just as the Mo_6X_8 units in the CHEVREL phases tolerate a certain lack of electrons (e.g. 20 instead of 24 skeleton electrons), clusters with M_6X_{12} units which have fewer than 16 skeleton electrons are also possible. For example, in Zr_6I_{12} there are only 12 skeleton electrons, and $Sc_7Cl_{12} = Sc^{3+}[Sc_6Cl_{12}]^{3-}$ has only nine.

Wade Clusters

K. WADE has put forward some rules that relate the composition of a cluster to the number of its valence electrons [95]. The rules were derived for boranes and, first of all, they apply to these, i.e. to compounds with a severe electron deficiency. To calculate the wave functions of a *closo* cluster with n atoms, the coordinate systems of all n atoms are oriented with their z axes radially to the center of the polyhedron. The contribution of the s orbitals can be estimated best by combining them with the p_z orbitals to form sp hybrid orbitals. One of the two sp orbitals of an atom points radially to the center of the cluster, the other one radially outwards. The latter is used for bonding with external atoms (e.g. with the H atoms of the $B_6H_6^{2-}$ ion). The n sp orbitals pointing inwards combine to give one bonding and $n-1$ nonbonding or antibonding orbitals. The orbitals p_x and p_y of every atom are oriented tangentially to the cluster and combine to give n bonding and n antibonding orbitals (Fig. 73). Altogether, we obtain $n+1$ bonding orbitals for the cluster skeleton. From this follows the WADE rule: *a stable closo cluster requires $2n+2$ skeleton electrons.* This is a lower number of electrons than that required for an electron precise cluster or for a cluster with $2e3c$ bonds, with one exception: a tetrahedral cluster with $2e3c$ bonds on its four faces requires only 8 electrons, whereas it should have 10 electrons according to the WADE rule; the WADE rule does not apply to tetrahedra. In fact, *closo*-boranes with the composition $B_nH_n^{2-}$ are known only for $n \geq 5$.

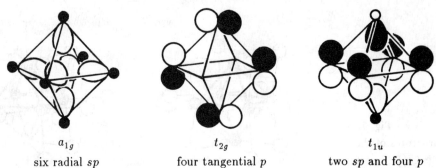

a_{1g}

six radial *sp*

t_{2g}

four tangential *p*

t_{1u}

two *sp* **and four** *p*

Fig. 73

Combinations of atomic orbitals that result in bonding molecular orbitals in an octahedral cluster such as $B_6H_6^{2-}$. For the triply degenerate orbitals t_{2g} and t_{1u} only one of each is plotted; for each of them, two further, equal orbitals exist with two other orientations

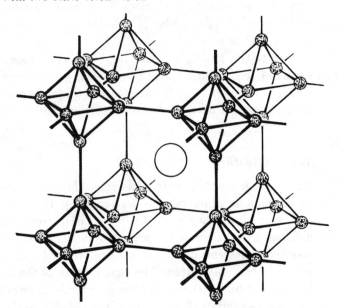

Fig. 74

The structure of CaB_6

The *closo*-boranes $B_nH_n^{2-}$ ($5 \leq n \leq 12$) and the carboranes $B_nC_2H_{n+2}$ are showpieces for the mentioned WADE rule. Further examples include the B_{12} icosahedra in elemental boron (Fig. 54) and certain borides such as CaB_6. In CaB_6, B_6 octahedra are linked with each other via normal $2e2c$ bonds (Fig. 74). Six electrons per octahedron are required for these bonds; together with the $2n + 2 = 14$ electrons for the octahedron skeleton this adds up to a total of 20 valence electrons. The boron atoms supply $3 \times 6 = 18$ of them, and calcium the remaining two.

WADE has stated some further rules for open clusters that are interpreted as polyhedra with missing vertices.* They are of importance for boranes. We do

---

Nido clusters: one missing polyhedron vertex, $n + 2$ bonding skeleton orbitals; *arachno* clusters: two missing vertices, $n + 3$ bonding skeleton orbitals.

not deal with them as they are discussed in detail in textbooks concerning the chemistry of the elements. WADE has also extended the application of his rules to transition metal clusters; the further extension by D.M.P. MINGOS under the name *polyhedron skeleton electron pair theory* (PSEPT) [96,97] mainly concerns the bonding in metal carbonyl and metal phosphine clusters, i.e. organometallic compounds; these are beyond the scope of this book.

Clusters with Interstitial Atoms

Clusters derived from metals which have only a few electrons can relieve their electron deficit by incorporating atoms inside. This is an option especially for octahedral clusters which are able to enclose a binding electron pair anyway. The interstitial atom usually contributes all of its valence electrons to the electron balance. Nonmetal atoms such as H, B, C, N, and Si as well as metal atoms such as Be, Al, Mn, Fe, Co, and Ir have been found as interstitial atoms.

Transition metals of the third and fourth group form many octahedral clusters that are isostructural to those of the less electron deficient elements of the following groups, but they contain additional atoms in their centers (Fig. 75). Starting from the above-mentioned Nb_6Cl_{14} (Fig. 72), we can substitute the niobium atoms for zirconium atoms; the number of available electrons is then reduced by six. This loss can partly be compensated by introducing a carbon atom in the Zr_6 octahedron. Despite the slightly inferior number of electrons the cluster in Zr_6CCl_{14} is stable due to some changes in the bonding. The more electronegative atom in the center of the cluster pulls electron density inwards, thus weakening the Zr–Zr bonds to some extent, but stronger bonding interactions with the C atom emerge.

On the other hand, the metal–metal bonds are strengthened when the interstitial atom is a metal atom. Nb_6F_{15}, for instance, consists of Nb_6F_{12} clusters of the same kind as in the $Nb_6Cl_{12}^{2+}$ unit; they are linked by all six of their vertices via bridging fluorine atoms, forming a network. Th_6FeBr_{15} has the same kind of structure, but with an additional Fe atom in the octahedron center (Fig. 75).

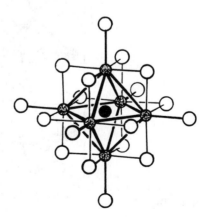

Fig. 75
Cluster unit with an interstitial atom in compounds such as Zr_6CCl_{14} and Th_6FeBr_{15}

Nb_6F_{15} has one electron less than required for the eight $2e3c$ bonds; in Th_6Br_{15} a further six electrons are missing. The intercalated Fe atom (d^8) supplies these seven electrons; the eighth electron remains with the Fe atom.

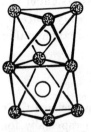

Even the extremely electron deficient alkali metals can form clusters when interstitial atoms contribute to their stabilization. Compounds of this kind are the alkali suboxides such as Rb_9O_2; it has two octahedra sharing a common face, and each one is occupied by one O atom. However, the electron deficiency is so severe that metallic bonding is needed between the clusters. In a way, these compounds are metals, but not with single metal ions as in the pure metal Rb^+e^-, but with a constitution $[Rb_9O_2]^{5+}(e^-)_5$, essentially with ionic bonding in the cluster.

cluster in Rb_9O_2

Condensed Clusters

Another possibility for relieving the electron deficiency consists of joining clusters to form larger building blocks. Such compounds have recently been studied intensively by A. SIMON [112,113,114] and J.D. CORBETT [115]. Among the known condensed clusters the majority consist of M_6 octahedra linked with each other. When joining M_6X_8 or M_6X_{12} units in such a way that metal atoms "merge" with one another, some of the X atoms have to be "merged" also.

Fig. 76 shows a possibility for the condensation of M_6X_8 clusters. Merging *trans* vertices of octahedra to a linear chain requires that opposite faces of the X_8 cubes also merge; every X atom is thus shared by two cubes. The resulting composition is M_5X_4. The relative arrangement of chains bundled in parallel allows the coordination of X atoms of one chain to the octahedra vertices of four adjacent chains, in a similar way as in the CHEVREL phases. Compounds with

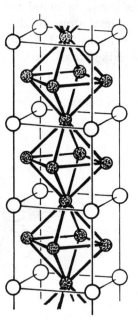

Fig. 76
Condensed M_6X_8 clusters in Ti_5Te_4

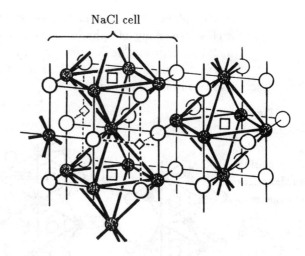

Fig. 77

Condensed M_6X_{12} clusters forming chains in TiO and the interpretation as an NaCl structure with vacancies. $\Diamond$ = vacant cation site, $\Box$ = vacant anion site

this structure are known with M = Ti, V, Nb, Ta, Mo and X = S, Se, Te, As, Sb, e.g. Ti_5Te_4. They have 12 (Ti_5Te_4) to 18 (Mo_5As_4) skeleton electrons per octahedron. Eight of the electrons form four $2e2c$ bonds at the four equatorial edges of the octahedron; the remaining electrons are oriented along the other octahedron edges, and their interaction in the chain direction results in metallic energy bands.

The corresponding condensation of M_6X_{12} clusters to chains and their parallel assembly as in Fig. 77 results in the TiO structure. The octahedra in a chain share vertices. Eight of the twelve X atoms on top of the octahedron edges are merged with eight others from the next clusters in the chain. Furthermore, half of them simultaneously belong to clusters of adjacent chains. These eight atoms therefore contribute to one cluster with fractions of $4/2 + 4/3$. Of the remaining four X atoms two simultaneously are part of one further cluster each and two simultaneously belong to two other clusters in neighboring chains; their contributions to one cluster are $2/2$ and $2/3$. Therefore, the general composition is $M_4M_{2/2}X_{4/2}X_{4/3}X_{2/2}X_{2/3}$ or M_5X_5. The structure can also be described as a defect NaCl structure. In NaCl, strings of alternate Na^+ and Cl^- ions are present; if we remove all Cl^- ions in one such string, void Na^+ octahedra sharing vertices in a chain remain. To balance the charges, we remove Na^+ ions in an adjacent string. As a result, one sixth of all Na^+ and Cl^- positions remains vacant. A similar arrangement, but with octahedra linked in three dimensions, is found in NbO (it has $\frac{1}{4}$ of the NaCl positions vacant).

Chains with the composition $M_2M_{4/2}X_{8/2} = M_4X_4$ are the result of the condensation of M_6X_8 clusters by merging opposite octahedron edges. They are known for lanthanoid halides like Gd_2Cl_3; they have additional halogen atoms placed in between the chains (Fig. 78). The cluster condensation can be carried on: the chains of octahedra sharing edges can be joined to double-strands and finally to layers of octahedra (Fig. 78). Every layer consists of metal atoms in two planes arranged in the same way as two adjacent layers of atoms in a closest-packing of spheres. This is simply a two-dimensional section from a metal structure. The X atoms occupy positions between the metal layers and act as "insulating" layers. Substances like ZrCl that have this structure behave like two-dimensional metals.

Fig. 78
Condensation of M_6X_8 clusters by sharing octahedra edges to yield chains in Gd_2Cl_3, double-strands in Sc_7Cl_{10} and layers in ZrCl. Chlorine atoms are on top and below of the triangular octahedron faces

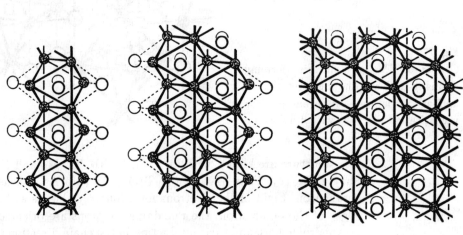

12.5   Problems

12.1 Use the extended $8-N$ rule to decide whether the following compounds are polyanionic, polycationic or simple ionic.

(a) Be_2C; (b) Mg_2C_3; (c) ThC_2; (d) Li_2Si; (e) In_4Se_3; (f) KSb; (g) Nb_3Cl_8; (h) TiS_2.

12.2 Which of the following compounds should be ZINTL phases?

(a) Y_5Si_3; (b) CaSi; (c) CaO; (d) K_3As_7; (e) NbF_4; (f) $LaNi_5$.

12.3 Draw valence bond formulas for the following ZINTL anions.

(a) $Al_2Te_6^{6-}$; (b) $[SnSb_3^{5-}]_\infty$; (c) $[SnSb^-]_\infty$; (d) $[Si^{2-}]_\infty$; (e) P_2^{4-}.

12.4 State which of the following clusters is electron precise, may have $2e3c$ bonds or fulfils the WADE rule for *closo* clusters.

(a) $B_{10}C_2H_{12}$ (icosahedron); (b) $Re_6(\mu_3\text{-}S)_4(\mu_3\text{-}Cl)_4\mu_1\text{-}Cl_6$ (octahedron);

(c) $Pt_4(\mu_3\text{-}H)_4(\mu_1\text{-}H)_4(PR_3)_4$ (tetrahedron); (d) $Os_5(CO)_{16}$ (trigonal bipyramid);

(e) $Rh_6(CO)_{16}$ (octahedron).

13 Sphere Packings. Metal Structures

Metals are electron deficient materials in which atoms are held together by multicenter bonds. The entire set of atoms in a crystal contributes to the multicenter bonds; the valence electrons are delocalized throughout the crystal; more details are given in chapter 9. The bonding forces act evenly on all atoms: usually there are no prevalent local forces that cause some specific coordination geometry around an atom in such a way as in a molecule. In what way the atoms arrange themselves in a metallic crystal depends in the first place on how a most dense packing can be achieved geometrically. However, in the second place the electronic configuration and the valence electron concentration do have some influence; they determine which of several possible packing variants will actually occur. In principle, band structure calculations should allow us to differentiate these variants. However, the energy differences between various packing types can be smaller than the error inherent in (approximate) calculations, thus impeding reliable predictions.

If atoms are considered as hard spheres, the packing density can be expressed by the space filling SF of the spheres. It is:

$$SF = \frac{4\pi}{3V} \sum_i Z_i r_i^3 \tag{27}$$

V = volume of the unit cell
r_i = radius of the i-th kind of sphere
Z_i = number of spheres of the i-th kind in the unit cell

If only one kind of sphere is present and all dimensions are referred to the diameter of one sphere, i.e. if we set the diameter to be 1 and the radius to be $r = \frac{1}{2}$, we obtain:

$$SF = \frac{\pi}{6} \cdot \frac{Z}{V} = 0.5236 \frac{Z}{V}$$

13.1   Closest Packings of Spheres

In order to fill space in the most economic way with spheres of equal size, we arrange them in a closest packing of spheres.* The closest arrangement of spheres in a plane is a *hexagonal layer* of spheres (Fig. 79). In such a layer every sphere has six adjacent spheres; six voids remain between a sphere and its six adjacent spheres. The distance from one void to the *next but one* void is exactly as long as the distance between the centers of two adjacent spheres. Let us mark the position of the sphere centers by A as in Fig. 79, and the positions of the voids by B and C. The closest stacking of layers requires that a layer in the position A be followed by a layer having its spheres in hollows on top of either the voids B or the voids C. Generally, there exist three possible layer positions in a closest stack of hexagonal layers; a layer can only be followed by another layer in a different position (A cannot be followed by A etc.).

Fig. 79

Arrangement of
spheres in a
hexagonal layer
and the relative
position of the
layer positions A,
B and C

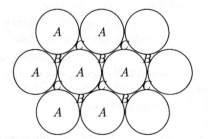
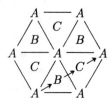

The layer sequence $ABCABC\ldots$ is marked by arrows in Fig. 79. In this sequence all arrows point in the same direction. In a sequence ABA one arrow would point in one direction, and the next arrow would point in the opposite direction. If we designate the direction $A \to B = B \to C = C \to A$ by $+$ and $A \gets B = B \gets C = C \gets A$ by $-$, we can characterize the stacking sequence by a sequence of $+$ and $-$ signs (HÄGG, 1943). The symbolism can be abbreviated according to ZHDANOV by a sequence of numbers, with every number specifying how many equal signs are side by side; only the numbers of one periodically repeating unit are given. Another frequently used symbolism is that by JAGODZINSKI: a layer having its two adjacent layers in different positions (e.g. the layer B in the sequence ABC), is designated by c (c for cubic); if its two adjacent layers have the same position (e.g. B in the sequence ABA), the symbol is h (for hexagonal).

Although the number of possible stacking sequences is infinitely large, predominantly only the following two are observed:

---

*That no sphere packing can have a higher density than the closest packings of spheres described here was not proven until 1991 by W.Y. Hsiang. Spheres surrounding a central sphere icosahedrally are not in contact with one another, i.e. there is a little bit more space than for twelve neighboring spheres. However, icosahedra cannot be packed in a space-filling manner. Some nonperiodic packings of spheres have been described that have densities close to the density of closest packings of spheres.

	cubic closest-packing (c.c.p.)	hexagonal closest-packing (h.c.p.)
stacking sequence	$...ABCABC...$	$...ABABAB...$
HÄGG symbol	$...++++++...$	$...+-+-+-...$
ZHDANOV symbol	∞	11
JAGODZINSKI symbol	c	h

Cubic closest-packing is also called **c**opper type, hexagonal closest-packing is also called **m**agnesium type. In the cubic closest-packing the spheres have a face-centered cubic (f.c.c.) arrangement (Fig. 80); the stacking direction of the hexagonal layers is perpendicular to either of the body diagonals across the cube. The coordination number of every sphere is 12 for both sphere packings. The coordination polyhedron is a cuboctahedron for cubic closest-packing; a cuboctahedron can be regarded either as a truncated cube or as a truncated octahedron (cf. Fig. 2, p. 5). The coordination polyhedron for hexagonal closest-packing is an anticuboctahedron; it results when two opposite triangular faces of a cuboctahedron are mutually rotated by 30°.

More complicated stacking sequences occur less frequently. Some have been observed among the lanthanoids:

	stacking sequence	JAGODZINSKI	ZHDANOV
La, Pr, Nd, Pm	$...ABAC...$	hc	22
Sm	$...ABACACBCB...$	hhc	21

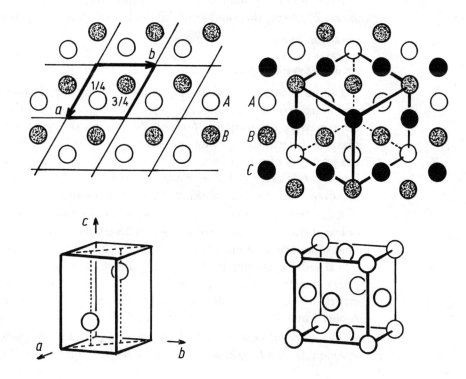

Fig. 80
Unit cells for hexagonal (left) and cubic closest-packing of spheres. Top row: projections in the stacking direction. Spheres are drawn smaller than their actual size. Spheres with the same coloring form hexagonal layers as in Fig. 79

Gadolinium to thulium as well as lutetium form hexagonal closest-packings of spheres. The proportion of h layers hence increases with increasing number of f electrons. The electronic configuration controls the kind of packing adopted; the influence of the $4f$ shell decreases with the atomic number. Being in the interior of an atom, with increasing nuclear charge the f shell experiences a stronger contraction than the $5d$ and the $6s$ shells, i.e. the lanthanoid contraction shows up to a higher degree inside the atoms than can be seen in the atomic radii. The influence of the f electrons also expresses itself in the behaviour of the lanthanoids under pressure; when compressed, the outer shells are squeezed more than the inner ones, and the f electrons gain influence resulting in structures with more c layers:

	normal pressure	high pressure
Pr, Nd	hc	c
Sm	hhc	hc
Gd, Tb, Dy, Ho	h	hhc

Finally, the influence of the electronic configuration also shows up in the exceptions: europium and ytterbium, having $4f$ shells "prematurely" half and completely filled, respectively, have structures which do not follow the sequence of the other metals (Table 18, p. 145; configuration for Eu $4f^7 6s^2$ instead of $4f^6 5d^1 6s^2$, for Yb $4f^{14} 6s^2$ instead of $4f^{13} 5d^1 6s^2$. These elements also have irregular atomic radii, cf. Tab. 3, p. 32). There is also an irregularity at the beginning of the series, since cerium adopts a cubic closest-packing.

With increasing numbers of hexagonal layers in one periodically repeating packet of layers, the number of different possible stacking variants increases:

number of layers per layer packet:	2	3	4	5	6	7	8	9	10	11	12	20
number of stacking variants:	1	1	1	1	2	3	6	7	16	21	43	4625

The notable predominance of simple stacking forms is an expression of the *symmetry principle: among several feasible structure types those having the highest symmetry are normally favored.* We discuss the reasons for and the importance of this principle in more detail in section 19.2. The frequent observance of the *principle of most economic filling of space*, i.e. of purely geometric aspects, is also remarkable: of the 94 elements with known structures in the solid state, 53 adopt closest sphere packings at normal conditions.

Aside from the ordered stacking sequences we have considered so far, a more or less statistical sequence of hexagonal layers can also occur. Since there is some kind of an ordering principle on the one hand, but on the other hand the periodical order is missing in the stacking direction, this is called an *order-disorder* (OD) structure. In this particular case, it is a *one-dimensionally disordered* structure, since the order is missing only in one dimension. When cobalt is cooled from 500 °C it exhibits this kind of disorder.

The *space filling* is the same for all stacking variants of closest sphere packings. It amounts to $\pi/(3\sqrt{2}) = 0.7405$ or 74.05 %.

Structures that can be considered as stacking variants of layers such as the closest packings of spheres are called *polytypes*.

13.2 The Body-centered Cubic Packing of Spheres

The space filling in the body-centered cubic packing of spheres is less than in the closest sphere packings, but the difference is moderate. The fraction of space filled amounts to $\frac{1}{8}\pi\sqrt{3} = 0.6802$ or 68.02 %. The reduction of the coordination number from 12 to 8 seems to be more serious; however, the difference is actually not so serious because in addition to the 8 directly adjacent spheres every sphere has 6 further neighbors that are only 15.5 % more distant (Fig. 81). The coordination number can be designated by 8+6.

Corresponding to its inferior space filling, the body-centered cubic packing of spheres is less frequent among the element structures. None the less, 15 elements crystallize with this structure. As tungsten is one of them, the term tungsten type is sometimes used for this kind of packing.

These 15 elements have this structures at normal conditions. If we also include the modifications adopted by some elements at higher temperatures or pressures, we obtain the following statistics:

Of the 94 elements with known structures, 68 crystallize with one of the sphere packings discussed so far (cf. Table 18). These 68 elements exhibit a total of about 115 modifications; they are distributed as follows:

	normal structure	total
closest packings	53	80
body-centered cubic	15	31
other structures		4
	68	115

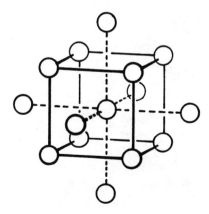

Fig. 81
Unit cell of the body-centered cubic sphere packing and the coordination geometry around one sphere

Most of the additional modifications are high temperature forms. As the figures show, the body-centered cubic packing of spheres gains importance at higher temperatures. This is in accordance with the GOLDSCHMIDT rule:

Increased temperatures favor structures with lower coordination numbers.

The body-centered cubic packing is also adopted by the high temperature modifications of some compounds that consist of molecules rotating in these structures and therefore have a spherical shape on average, such as MoF_6.

13.3   Other Metal Structures

As can be seen in Table 18, the solid state structures of most metals correspond to one of the sphere packings discussed (in some cases with certain distortions). Some metals, however, show structure types of their own: Ga, Sn, Bi, Po, Mn, Pa, U, Np, and Pu. For Sn, Bi, and Po refer to pp. 110, 99 and 97, respectively. Gallium has a rather unusual structure in which every Ga atom has coordination number 1+6; one of the seven adjacent atoms is significantly closer than the others ($1\times$ 244 pm, $6\times$ 270 to 279 pm). This has been interpreted as a metal structure consisting not of single atoms, but of Ga–Ga pairs with a covalent bond. The notably low melting point of gallium (29.8°C) shows this structure to be not especially stable; it seems to be only an "expedient". There also seems to be no optimal structure for Mn, U, Np, and Pu, as these elements form remarkably many polymorphous forms with rather peculiar structures. For example, the unit cell of α-Mn, the modification stable at room temperature, contains 58 atoms with four different kinds of coordination polyhedra having coordination numbers 12, 13, and 16.

Table 18 The element structures of the metals at ambient conditions

h = hexagonal closest-packing

c = cubic closest-packing

hc, hhc = other stacking varieties of closest-packing

i = body-centered cubic packing

⋈ = structure type of its own

* = distorted

The solid noble gases also adopt closest packings of spheres: He h, Ne...Xe c

Li i	Be h												
Na i	Mg h										Al c		
K i	Ca c	Sc h	Ti h	V i	Cr i	Mn ⋈	Fe i	Co h	Ni c	Cu c	Zn $h^\star$	Ga ⋈	
Rb i	Sr c	Y h	Zr h	Nb i	Mo i	Tc h	Ru h	Rh c	Pd c	Ag c	Cd $h^\star$	In $c^\star$	Sn ⋈
Cs i	Ba i	La hc	Hf h	Ta i	W i	Re h	Os h	Ir c	Pt c	Au c	Hg $c^\star$	Tl h	Pb c
Fr	Ra i	Ac c											

Ce c	Pr hc	Nd hc	Pm hc	Sm hhc	Eu i	Gd h	Tb h	Dy h	Ho h	Er h	Tm h	Yb c	Lu h
Th c	Pa ⋈	U ⋈	Np ⋈	Pu ⋈	Am hc	Cm hc	Bk c, hc	Cf	Es	Fm	Md	No	Lr

13.4   Problems

13.1 State the JAGONDZINSKI and the ZHDANOV symbols for the closest packings of spheres with the following stacking sequences:

(a) $ABABC$; (b) $ABABACAC$.

13.2 State the stacking sequence (by A, B and C) for the closest packings of spheres with the following JAGONDZINSKI or ZHDANOV symbols:

(a) hcc; (b) $cchh$; (c) 221.

14 The Sphere Packing Principle for Compounds

The geometric principles for the packing of spheres do not only apply for pure elements. As might be expected, the sphere packings discussed in the preceding chapter are also frequently encountered when similar atoms are combined, especially among the numerous intermetallic compounds. Furthermore, the same principles also apply for many compounds consisting of elements which differ widely.

14.1 Ordered and Disordered Alloys

Different metals can very frequently be mixed with each other in the molten state, i.e. they form homogeneous solutions. A solid solution is obtained by quenching the liquid; in the *disordered alloy* obtained this way, the atoms are distributed randomly. When cooled slowly, in some cases solid solutions can also be obtained. However, it is more common that a segregation takes place, in one of the following ways:

1. The metals crystallize separately (complete segregation).

2. Two kinds of solid solutions crystallize, a solution of metal 1 in metal 2 and vice versa (limited miscibility).

3. An alloy with a specific composition crystallizes; its composition may differ from that of the liquid (formation of an intermetallic compound). The composition of the liquid can change during the crystallization process and further intermetallic compounds with other compositions may crystallize.

The phase diagram shows which of these possibilities applies and whether intermetallic compounds will eventually form (cf. section 3.2, p. 15).

The tendency to form solid solutions depends mainly on two factors, namely the chemical relationship between the metals and the relative size of their atoms.

Two metals that are chemically related and that have atoms of nearly the same size form disordered alloys with each other. Silver and gold, both crystallizing with cubic closest-packing, have atoms of nearly equal size (radii 144.4 and 144.2 pm, respectively). They form solid solutions (mixed crystals) of arbitrary composition in which the silver and the gold atoms randomly occupy the positions of the sphere packing. Related metals, especially from the same group of the periodic table, generally form solid solutions which have any composition if their atomic radii do not differ by more than approximately 15% (e.g. Mo/W; K/Rb, K/Cs, but not Na/Cs). If the elements are less similar, there may be a limited miscibility as in the case of, for example, Zn in Cu (molar fraction of Zn maximally 38.4%) and Cu in Zn (maximally 2.3% Cu); copper and zinc additionally form intermetallic compounds (cf. section 14.4).

When the atoms differ in size or when the metals are chemically different, structures with ordered atomic distributions are considerably more likely. Since the transition from a disordered to an ordered state involves a decrease in entropy, and since the transition only takes place when $\Delta G = \Delta H - T\Delta S < 0$, the transformation enthalpy ΔH must be negative. The ordered structure therefore is favored energetically; it has a larger lattice energy.

An ordered distribution of spheres of different sizes always allows a better space filling; the atoms are closer together, and the attractive bonding forces become more effective. As for the structures of other types of compound, we observe the validity of the *principle of the most efficient filling of space*. A definite order of atoms requires a definite chemical composition. Therefore, metal atoms having different radii preferentially will combine in the solid state with a definite stoichiometry: they will form an intermetallic compound.

Even when complete miscibility is possible in the solid state, ordered structures will be favored at suitable compositions if the atoms have different sizes. For example: copper atoms are smaller than gold atoms (radii 127.8 and 144.2 pm, respectively); copper and gold form mixed crystals of any composition, but ordered alloys are formed with the stoichiometric ratios AuCu and AuCu$_3$ (Fig. 82). The degree of order is temperature dependent; with increasing temperatures the order decreases continuously. Therefore, there is no phase transition with a well-defined transition temperature. This can be seen in the temperature dependence of the specific heat (Fig. 83). Because of the form of the curve, this kind of order–disorder transformation is also called a Λ type transformation; it is observed in many solid state transformations.

14.2 Compounds with Close-packed Atoms

As in ionic compounds, the atoms in a binary intermetallic compound show a tendency, albeit less pronounced, to be surrounded by atoms of the other kind as far as possible. However, it is not possible to fulfil this condition simultaneously for both kinds of atoms if they form a closest-packed arrangement. For

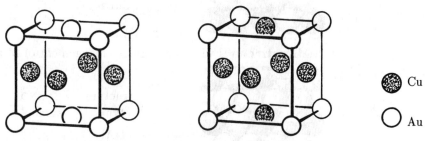

Fig. 82

The structures of the ordered alloys AuCu and AuCu$_3$. At higher temperatures they are transformed to alloys which have all atomic positions statistically occupied by the Cu and Au atoms

C_p /Jg^{-1}K^{-1}

0.75 –

0.50 –

0.25 –

100   200   300   400   500 T /°C

Fig. 83
Variation of the
specific heat C_p
with temperature
of AuCu$_3$
(Λ-type
transformation)

compositions MX$_n$ with $n < 3$ it cannot be fulfilled for either the M or the X atoms: in every case every atom has to have some adjacent atoms of the same kind. Only with a higher content of X atoms, begining with MX$_3$ ($n \geq 3$), are atomic arrangements possible in which every M atom is surrounded solely by X atoms; the X atoms, however, must continue to have other X atoms as neighbors.

Usually, the stoichiometry of the compound is fulfilled in every one of the hexagonal layers. This facilitates a rational classification of the extensive data: it is only necessary to draw a sketch of the atomic arrangement in one layer and to specify the stacking sequence (ZHDANOV or JAGODZINSKI symbol). The most important structure types of this kind are:

1. MX$_3$ structures with hexagonal arrangement of M atoms in one layer (M atoms gray in the figure, positions of M atoms in the following layer marked by $*$)

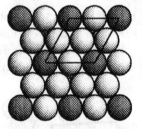

structure type	ZHDANOV symbol	JAGODZINSKI symbol
AuCu$_3$	∞	c
SnNi$_3$	11	h

2. MX$_3$ structures with rectangular arrangement of M atoms

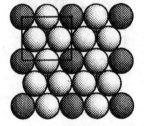

structure type	ZHDANOV symbol	JAGODZINSKI symbol
TiAl$_3$	∞	c
TiCu$_3$	11	h

3. MX structures with alternating strings of equal atoms

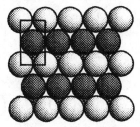

structure type	ZHDANOV symbol	JAGODZINSKI symbol
AuCu	∞	*c*
AuCd	11	*h*
TaRh	33	*hcc*

Because strings of the same atoms come to be adjacent when these layers are stacked, alternating layers of atoms of one kind each are formed. The new layers are planar in AuCu, and are inclined relative to the plane of the paper; in the unit cell (Fig. 82) they are parallel to the base plane. The layers of equal atoms are undulated in the other two structure types.

14.3 Structures Derived of Body-centered Cubic Packing (CsCl Type)

Disordered alloys may form when two metals are mixed if both have body-centered cubic structures and if their atomic radii do not differ by much (e.g. K and Rb). The formation of ordered alloys, however, is usually favored (at higher temperatures the tendency towards disordered structures increases). Even when metals which do not crystallize body-centered cubic themselves are combined in the appropriate stoichiometry, they can adopt such an arrangement. β-brass (CuZn) is an example; below 300 °C it has CsCl structure, but between 300 °C and 500 °C a Λ type transformation takes place resulting in a disordered alloy with body-centered cubic structure.

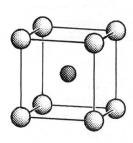

The **CsCl type** offers the simplest way to combine atoms of two different elements in the same arrangement as in body-centered cubic sphere packing: the atom in the center of the unit cell is surrounded by eight atoms of the other element in the vertices of the unit cell.* This way every atom only has adjacent atoms of the other element. This is a condition that cannot be fulfilled in a closest-packing of spheres (cf. preceding section).

Although the space filling of the body-centered cubic sphere packing is somewhat inferior to that of a closest-packing, the CsCl type thus turns out to be excellently suited for compounds with a 1:1 composition.

We presented the CsCl type in chapter 6 as an important structure type for ionic compounds. Its importance, however, is by no means restricted to this class of compounds: only about 12 out of 150 compounds with this structure are salt-like (e.g. CsI, TlBr), though at higher temperatures or higher pressures there are some 15 more (e.g. NaCl, KCl at high pressure; TlCN at high temperature with rotating CN⁻ ions). More than 130 representatives are intermetallic compounds, e.g. MgAg, CaHg, AlFe, and CuZn.

---

*The CsCl type itself is *not* body-centered, since a centered lattice requires atoms of the same kind in the vertices and the center of the unit cell.

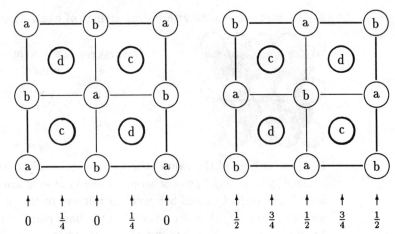

height: 0 $\frac{1}{4}$ 0 $\frac{1}{4}$ 0 $\frac{1}{2}$ $\frac{3}{4}$ $\frac{1}{2}$ $\frac{3}{4}$ $\frac{1}{2}$

Fig. 84
Superstructure of the CsCl type with eightfold unit cell. Left, lower half, right, upper half of the cell in projection onto the plane of the paper. a, b, c, and d designate four different kinds of atomic sites that can be occupied in the following ways:

a	b	c	d	structure type	examples
Al	Fe	Fe	Fe	Fe$_3$Al (Li$_3$Bi)	Fe$_3$Si
Al	Mn	Cu	Cu	MnCu$_2$Al (Heusler alloy)	
Tl	Na	Tl	Na	NaTl (Zintl phase)	LiAl, LiZn
Ag	Li	Sb	Li	Li$_2$AgSb (Zintl phase)	
As	□	Mg	Ag	MgAgAs (Zintl phase)	LiMgAs
Ca	□	F	F	CaF$_2$ (fluorite)	CuF$_2$, BaCl$_2$, ThO$_2$, TiH$_2$, Li$_2$O, Be$_2$C, Mg$_2$Sn
Zn	□	S	□	ZnS (sphalerite)	SiC, AlP, GaAs, CuCl
C	□	C	□	diamond	Si, α-Sn
Na	Cl	□	□	NaCl	LiH, AgF, MgO, TiC

Superstructures of the CsCl type result when the unit cell of the CsCl structure is multiplied and the atomic positions are occupied by different kinds of atoms. If we double the cell edges in all three dimensions, we obtain a cell that consists of eight subcells, each of which contains one atom in its center (Fig. 84). The 16 atoms in the cell can be subdivided into four groups of four atoms each; each group has a face-centered arrangement. Depending on how we distribute atoms of different elements among these four groups, we obtain different structure types, as listed in Fig. 84. The list includes possibilities with certain vacant atomic positions (marked with the SCHOTTKY symbol □ in the table). This option reduces the space filling; however, as long as the positions a and b are occupied by different kinds of atoms than the positions c and d, every atom still has only atoms of a different kind as nearest neighbors. Consequently, the structure types are adequate for ionic compounds, including ZINTL phases with simple "anions" such as As^{3-}, Sb^{3-} or Ge^{4-}.

The following series shows that the mentioned structure types are adopted by all kinds of compounds from purely ionic to purely metallic:

fluorite type and variants					Fe$_3$Al type and variants			W type
F$_2$Ca	Li$_2$O	Li$_2$Te	LiMgAs	Mg$_2$Sn	Cu$_3$Sb	Cu$_2$MnAl	Fe$_3$Al	Fe
ionic				$\longrightarrow$				metallic

When covalent bonds favor neighbors of the same element, the positions c and d can also be occupied by atoms of the same kind as in a or b. This applies for diamond and for the ZINTL phase NaTl, which can be regarded as a network of Tl$^-$ particles with a diamond structure that encloses Na$^+$ ions (cf. Fig. 65, p. 123).

14.4 Hume–Rothery Phases

HUME–ROTHERY phases (brass phases, "electron compounds") are certain alloys with the structures of the different types of brass (brass = Cu–Zn alloys). They are classical examples of the structure-determining influence of the valence electron concentration (VEC) in metals. VEC = (number of valence electrons)/(number of atoms). A survey is given in Table 19.

α-brass is a solid solution of zinc in copper which has the structure of copper; the atoms statistically occupy the positions of the cubic closest-packing of spheres. In β-brass, which is obtained by quenching the melt, the atoms also have a random distribution, and the packing is body-centered cubic. The composition is not exactly CuZn; this phase is stable only if the fraction of zinc atoms amounts to 45 to 48 %. The γ phase also has a certain range of compositions from Cu$_5$Zn$_{6.9}$ to Cu$_5$Zn$_{9.7}$. The γ-brass structure can be described as a superstructure of the body-centered cubic packing with tripled lattice constants, so that the unit cell has a volume enlarged by a factor $3^3 = 27$. However, the cell only contains 52 instead of $2 \times 27 = 54$ atoms; there are two vacancies. The distribution of the vacancies is ordered; there are four kinds of positions for the metal atoms in a ratio of 3:2:2:6, but a random distribution may occur to some extent (in Cu$_5$Zn$_8$ the distribution is 3Cu:2Cu:2Zn:6Zn). A brass sample

Table 19 Brass phases

	composition	VEC	structure type	examples
α	Cu$_{1-x}$Zn$_x$, x=0 to 0.38	1 to 1.38	Cu	
β	CuZn	$1.50 = 21/14$	W	AgZn, Cu$_3$Al, Cu$_5$Sn
γ	Cu$_5$Zn$_8$	$1.62 = 21/13$	Cu$_5$Zn$_8$	Ag$_5$Zn$_8$, Cu$_9$Al$_4$, Na$_{31}$Pb$_8$
ε	CuZn$_3$	$1.75 = 21/12$	Mg	AgZn$_3$, Cu$_3$Sn, Ag$_5$Al$_3$
η	Cu$_x$Zn$_{1-x}$, x=0 to 0.02	1.98 to 2	Mg	

with a composition outside of the mentioned ranges consists of a mixture of the two neighboring phases.

Because of the permitted composition ranges, alloys with rather different compositions can adopt the same structure, as can be seen by the examples in Table 19. The determining factor is the valence electron concentration, which can be calculated as follows:

$$
\begin{array}{lll}
\text{AgZn} & \frac{1+2}{2} = \frac{3}{2} = \frac{21}{14} & \text{Ag}_5\text{Zn}_8 \quad \frac{5+16}{13} = \frac{21}{13} \quad \text{AgZn}_3 \quad \frac{1+6}{4} = \frac{7}{4} = \frac{21}{12} \\[2mm]
\text{Cu}_3\text{Al} & \frac{3+3}{4} = \frac{6}{4} = \frac{21}{14} & \text{Cu}_9\text{Al}_4 \quad \frac{9+12}{13} = \frac{21}{13} \quad \text{Cu}_3\text{Sn} \quad \frac{3+4}{4} = \frac{7}{4} = \frac{21}{12} \\[2mm]
\text{Cu}_5\text{Sn} & \frac{5+4}{6} = \frac{9}{6} = \frac{21}{14} & \text{Na}_{31}\text{Pb}_8 \quad \frac{31+32}{39} = \frac{21}{13} \quad \text{Ag}_5\text{Al}_3 \quad \frac{5+9}{8} = \frac{14}{8} = \frac{21}{12}
\end{array}
$$

The theoretical interpretation relating the valence electron concentration and the structure has been put forward by H. JONES [41]. If we start from copper and add more and more zinc, the VEC increases. The added electrons have to occupy higher energy levels, i.e. the energy of the FERMI limit is raised and comes closer to the limits of the first BRILLOUIN zone. This is approached at about VEC = 1.36. Higher values of the VEC require the occupation of antibonding states; now the body-centered cubic lattice becomes more favorable as it allows a higher VEC within the first BRILLOUIN zone, up to approximately VEC = 1.48.

14.5   Laves Phases

The term LAVES phases is used for certain alloys with the composition MM'_2, the M atoms being bigger than the M′ atoms. The classical representative is $MgCu_2$; its structure is shown in Fig. 85. It can be regarded as a superstructure of the CsCl type as in Fig. 84, with the following occupation of the positions a, b, c, and d:

$$\text{a: Mg} \qquad \text{b: Cu}_4 \qquad \text{c: Mg} \qquad \text{d: Cu}_4$$

We thus have placed a tetrahedron of four Cu atoms instead of a single atom in the position b; the same kind of Cu tetrahedron then also results at the position d. The magnesium atoms by themselves have the same arrangement as in diamond.

In addition to this cubic LAVES phase, a variant with magnesium atoms arranged as in hexagonal diamond exists in the $MgZn_2$ type, and further polytypes are known.

The copper atoms of the $MgCu_2$ type are linked to a three-dimensional network of vertex-sharing tetrahedra (Fig. 85), so that every Cu atom is linked with six other Cu atoms. If we assume an electron distribution according to the formula $Mg^{2+}(Cu^-)_2$, every copper atom attains a valence electron concentration of $\text{VEC(Cu)} = \frac{1}{2}(1 \cdot 2 + 2 \cdot 11) = 12$. Taking equation (19) from p. 117, adapted to transition metals as $b(X) = 18 - \text{VEC(M)}$, we calculate $b(X) = 18 - 12 = 6$ bonds per Cu atom. In other words, by linking Cu^- particles this way, copper attains the electron configuration of the next noble gas. $MgCu_2$, in a way, fulfils the rules for a ZINTL phase. Nevertheless, LAVES phases customarily are not considered to be ZINTL phases; some 170 intermetallic compounds

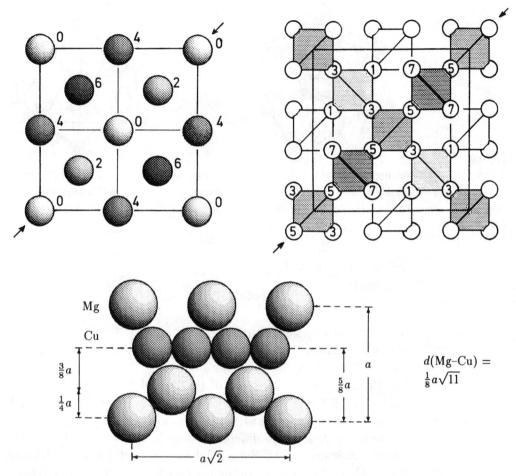

Fig. 85

Structure of the LAVES phase $MgCu_2$. Left: Mg partial structure. Right: Cu partial structure consisting of vertex-sharing tetrahedra. Numbers designate the height in the unit cell as multiples of $\frac{1}{8}$. Bottom: section across the cell in the diagonal direction marked by arrows in the top row, plotted with atomic radii corresponding to atoms in contact with each other (on a smaller scale than the upper part)

having the $MgCu_2$ structure are known, and most of them do not fulfil ZINTL's valence rule (e.g. $CaAl_2$, YCo_2, $LiPt_2$).

The space filling in the $MgCu_2$ type can be calculated with the aid of equation 27 (p. 139); the geometric relations follow from the bottom image in Fig. 85:
the four Cu spheres form a row along the diagonal of length $a\sqrt{2}$, therefore $r(Cu) = \frac{1}{8}\sqrt{2}\,a$;

two Mg spheres along the space diagonal of the unit cell are at a distance of $\frac{1}{4}\sqrt{3}\,a$, therefore $r(Mg) = \frac{1}{8}\sqrt{3}\,a$.
The ideal radius ratio is therefore

$$\frac{r(Mg)}{r(Cu)} = \sqrt{\frac{3}{2}} = 1.225$$

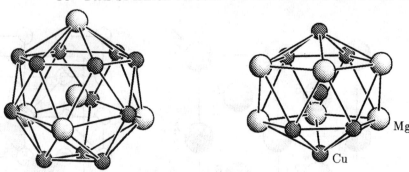

Fig. 86
FRANK–KASPER polyhedra in $MgCu_2$. The polyhedron around an Mg atom (c.n. 16) is composed of 12 Cu atoms and of four Mg atoms that form a tetrahedron by themselves; the Cu atoms form four triangles that are opposed to the Mg atoms. The polyhedron around a Cu atom (c.n. 12) is an icosahedron in which two opposite faces are occupied by Cu atoms

and the space filling is

$$\frac{4}{3}\pi\frac{1}{a^3}[8(\frac{1}{8}\sqrt{3}\,a)^3 + 16(\frac{1}{8}\sqrt{2}\,a)^3] = 0.710$$

(the unit cell contains 8 Mg and 16 Cu atoms). The space filling of 71.0 % is somewhat inferior than in a closest-packing of spheres (74.1 %). The coordination of the atoms is the following:
Mg: c.n. 16, 4 Mg at a distance of $\frac{1}{8}\sqrt{12}a$ and 12 Cu at a distance of $\frac{1}{8}\sqrt{11}a$;
Cu: c.n. 12, 6 Cu at a distance of $\frac{1}{8}\sqrt{8}a$ and 6 Mg at a distance of $\frac{1}{8}\sqrt{11}a$.

The coordination polyhedra are FRANK–KASPER polyhedra [71]. These are polyhedra with equal or different triangular faces, and at least five triangles meeting at every vertex. Such polyhedra allow for the coordination numbers 12, 14, 15, and 16. Fig. 86 shows the two FRANK–KASPER polyhedra occuring in $MgCu_2$. FRANK–KASPER polyhedra and the corresponding high coordination numbers are known among numerous intermetallic compounds.

The sketched model assuming hard spheres has a flaw: the sum of the atomic radii of Mg and Cu is smaller than the shortest distance between these atoms:

$$r(Mg) + r(Cu) \quad = \quad \tfrac{1}{8}(\sqrt{3}+\sqrt{2})a \quad = \quad 0.393\,a$$
$$d(Mg\text{–}Cu) \quad = \quad \tfrac{1}{8}\sqrt{11}\,a \quad = \quad 0.415\,a$$

Whereas the Mg atoms are in contact with each other and the Cu atoms are in contact with each other, the Cu partial structure "floats" inside of the Mg partial structure. The hard sphere model proves to be insufficient to account for the real situation: atoms are not really hard. The principle of the most efficient filling of space should rather be stated as the *principle of achieving the highest possible density*. Indeed, this shows up in the actual densities of the LAVES phases; they are greater than the densities of the components (in some cases up to 50% more). For example, the density of $MgCu_2$ is 5.75 $g\,cm^{-3}$, that is 7% more than the mean density of 5.37 $g\,cm^{-3}$ for 1 mole Mg + 2 moles Cu. Therefore, the atoms have a more dense packing in $MgCu_2$ than in the pure elements, the atoms are effectively smaller. According to the hard sphere model,

MgCu$_2$ should not be formed at all, as its space filling of 71% is inferior to that of both magnesium and copper, both of which crystallize with closest-packings of spheres (74% space filling).

It is mainly the Mg atoms that are affected by the compression of the atoms. The increase in density is the expression of a gain in lattice energy due to stronger bonding forces between the different kinds of atoms. These bonding forces have polar contributions since LAVES phases of this type experience a higher compression when the difference in the electronegativities of the atoms is higher. The polarity is an argument in favor of regarding LAVES phases in a similar way as ZINTL phases. More insight into the kind of bonding has been obtained by band structure calculations, which also allow the distinction of what electron counts favor the cubic MgCu$_2$ or the hexagonal MgZn$_2$ type [117].

14.6   Problems

14.1 Use Table 18 to decide whether the following pairs of metals are likely to form disordered alloys of arbitrary composition with each other.
(a) Mg/Ca; (b) Ca/Sr; (c) Sr/Ba; (d) La/Ac; (e) Ti/Mn; (f) Ru/Os; (g) Pr/Nd; (h) Eu/Gd.

14.2 Draw a section of each of the structure types presented on p. 148 corresponding to a plane running in the vertical direction of the figure and perpendicular to the plane of the paper.

14.3 What structure types result when the atomic positions in Fig. 84 are occupied in the following manner (A, B, C and D refer to chemical elements):
(a) a A, b A, c □, d B; (b) a A, b B, c C, d □; (c) a A, b A, c C, d D?

14.4 How is it possible that both Ag$_5$Zn$_8$ and Cu$_9$Al$_4$ have the γ-brass structure even though their stoichiometries differ?

14.5 Can an icosahedron be considered to be a FRANK–KASPER polyhedron?

15   Linked Polyhedra

The immediate surroundings of single atoms can be rationalized quite well with the aid of coordination polyhedra, at least when the polyhedra show a certain degree of symmetry to a good approximation. The most important polyhedra are presented in Fig. 2 (p. 5). Larger structural entities can be regarded as a system of linked polyhedra. Two polyhedra can be linked by sharing a common vertex, a common edge or a common face, i.e. they share one, two or three (or more) common "bridging" atoms (Fig. 3, p. 6).

Depending on the kind of polyhedron and the kind of linking, the resulting bond angles at the bridging atoms have a definite value or values confined within certain limits. The bond angle is fixed by geometry in the case of face-sharing polyhedra. By mutually rotating the polyhedra, the bond angle can be varied within certain limiting values for vertex-sharing and in some cases for edge-sharing polyhedra (Fig. 87; cf. also Fig. 104, p. 173). The values of the bond angles are listed in Table 20; they refer to undistorted tetrahedra and octahedra, and it is assumed that the closest contact of any two atoms corresponds to the distance of two adjacent atoms within a polyhedron. Distortions occur frequently and allow for an additional range of angles. Distortions may involve differing lengths of the polyhedron edges, but they may also come about by shifting the central atom out of the polyhedron center, thus changing the bond angles at the central atom and the bridging atoms even when the polyhedron edges remain constant.

Distortions of coordination polyhedra can often be interpreted according to the GILLESPIE–NYHOLM rules and by taking into account the electrostatic forces. For instance, a mutual repulsion of the Fe atoms can be perceived in the two edge sharing tetrahedra of the $(FeCl_3)_2$ molecule; it can be ascribed to their positive partial charges. The Fe–Cl distances to the bridging atoms thus

Table 20 Bond angles at the bridging atoms and distances between the central atoms M of linked tetrahedra and octahedra (disregarding possible distortions). The distances are given as multiples of the polyhedron edge length

		linking by		
		vertices	edges	faces
tetrahedron	bond angles	102.1 to 180°	66.0 to 70.5°	38.9°
	M–M distances	0.95 to 1.22	0.66 to 0.71	0.41
octahedron	bond angles	131.8 to 180°	90°	70.5°
	M–M distances	1.29 to 1.41	1.00	0.82

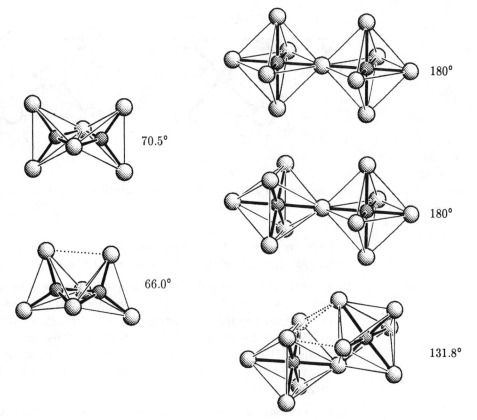

Fig. 87
Limits of the mutual rotation of vertex-sharing tetrahedra and of vertex-sharing octahedra and the resulting bond angles at the bridging atoms. The minimum distance between vertices of different polyhedra (dotted) was taken to be equal to the polyhedron edge

70.5°

180°

66.0°

180°

131.8°

become longer than the remaining Fe–Cl bonds. The Cl atoms adjust their positions by a slight deformation of the tetrahedra (Fig. 88). If the bridging atoms have a more negative partial charge than the terminal atoms, they counterbalance this kind of distortion since they exert a stronger attraction towards the central atoms which now only experience a decreased shift from the polyhedron centers. $(FeSCl_2)_2^{2-}$, which is isoelectronic to $(FeCl_3)_2$, is an example (Fig. 88; in order to compare the electrostatic forces, in a simplified manner one can assume the existence of ions Fe^{3+}, Cl^- and S^{2-}).

In what way polyhedra will join depends on several factors, which include:

1. Stoichiometry. Only very definite patterns of linking polyhedra are consistent with a given chemical composition.

2. The nature of the bridging atoms. They tend to attain certain bond angles and tolerate only bond angles within certain limits. Bridging sulfur, selenium, chlorine, bromine, and iodine atoms (having two lone electron pairs) favor angles close to 100°. This angle is compatible with vertex-sharing tetrahedra and with edge-sharing octahedra; however, examples with smaller angles among edge-sharing tetrahedra and face-sharing octa-

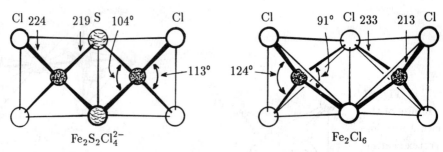

Fig. 88

Two edge sharing tetrahedra showing only minor distortions in the $Fe_2S_2Cl_4^{2-}$ ion and two more distorted tetrahedra in the Fe_2Cl_6 molecule. The distortions can be ascribed mainly to the electrostatic repulsion between the Fe atoms. Distances in pm

hedra are known. Bridging oxygen and fluorine atoms allow for angles up to 180°; frequently observed values are in the range 130 to 150°.

3. Bond polarity. Very polar bonds do not harmonize with edge-sharing poly-hedra and especially with face-sharing polyhedra because of the increased electrostatic repulsion between the central atoms (PAULING's third rule, p. 45). Central atoms in high oxidation states therefore favor vertex-sharing. If there are two kinds of central atom, those with the higher oxidation state will avoid having their polyhedra linked with one another (PAULING's fourth rule).

4. Interactions between the central atoms of the linked polyhedra. When a direct bond between the central atoms is advantageous, they tend to come close together. This favors edge-sharing or face-sharing arrangements. For example, the face-sharing of two oc-tahedra in the $[W_2Cl_9]^{3-}$ ion renders the formation of a W≡W triple bond possible; this way every tungsten atom gains elec-trons in addition to its electronic configuration d^3 and the elec-trons supplied by the ligands, thus attaining noble gas config-uration (18 valence electrons).

With our present knowledge, we often cannot understand, let alone pre-dict, the more profound details concerning the kind of linking. Why does BiF_5 form linear, polymeric chains, SbF_5 tetrameric molecules and AsF_5 monomeric molecules? Why are there chloro and not sulfur bridges in $(WSCl_4)_2$? Why does no modification of TiO_2 exist which has the quartz structure?

The composition of a compound is intimately related to the way of linking the polyhedra. An atom X with coordination number c.n.(X) that acts as a common vertex to this number of polyhedra, makes a contribution of $1/c.n.(X)$ to every polyhedron. If a polyhedron has n such atoms, this amounts to $n/c.n.(X)$ for this polyhedron. This can be expressed with NIGGLI formulae, as shown in the following sections. To specify the coordination polyhedra, the formalism presented at the end of section 2.1 and in Fig. 2 (p. 5) is useful.

15.1 Vertex-sharing Octahedra

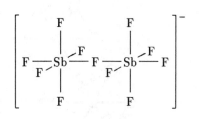

A single octahedral molecule has the composition MX_6. Two octahedra with a common vertex can be regarded as a unit MX_6 to which a unit MX_5 has been added, so that the composition is M_2X_{11}. If the addition of MX_5 units is continued, one obtains chain-like or ring-like molecules of composition $(MX_5)_n$ (Fig. 89). In these, every octahedron has four terminal atoms and two atoms that act as common vertices to other octahedra, corresponding to the NIGGLI formula $MX_{4/1}X_{2/2}$. If the two bridging vertices of every octahedron are mutually in *trans* position, the result is a chain; it can either be entirely straight as in BiF_5, with bond angles of 180° at the bridging atoms, or it can have a zigzag shape as in the CrF_5^{2-} ion, with bond angles between 132° and 180° (usually 132° to 150° for fluorides). If the two bridging vertices of every octahedron are in *cis* position, a large number of geometrical arrangements are possible. Among these, zigzag chains as in CrF_5 and tetrameric molecules as in $(NbF_5)_4$ are of importance. Again, the bond angles at the bridging atoms can have values from 132° to 180°. In pentafluorides, the frequent occurrence of angles having either 132° or 180° has to do with the packing of the molecules in the crystal: these two values result geometrically when the fluorine atoms for themselves form a hexagonal or a cubic closest-packing of spheres, respectively (this is discussed in more detail in chapter 16). The most important linking patterns for pentafluorides, pentafluoro anions and oxotetrahalides are:

	octahedron configuration	bond angle at bridging atom approx.	examples
rings $(MF_5)_4$	*cis*	180°	$(NbF_5)_4$, $(MoF_5)_4$
rings $(MF_5)_4$	*cis*	132°	$(RuF_5)_4$, $(RhF_5)_4$
linear chains	*trans*	180°	BiF_5, UF_5, $WOCl_4$
zigzag chains	*trans*	150°	$Ca[CrF_5]$, $Ca[MnF_5]$
zigzag chains	*cis*	180°	$Rb_2[CrF_5]$
zigzag chains	*cis*	152°	VF_5, CrF_5, $MoOF_4$

The layer shown in the left part of Fig. 90 represents the most important way to join octahedra by sharing four vertices each; the composition is $MX_4 = MX_{2/1}X_{4/2}$ or $M^{[o]}X_2^{[2l]}X_2$. Layers of this kind occur among some tetrafluorides such as SnF_4 and PbF_4 as well as in the anions of $Tl[AlF_4]$ and $K_2[NiF_4]$. The K_2NiF_4 type has been observed in a series of fluorides and oxides: K_2MF_4 with M = Mg, Zn, Co, Ni; Sr_2MO_4 with M = Sn, Ti, Mo, Mn, Ru, Rh, Ir, and some others. The preference of K^+ and Sr^{2+} has to do with the sizes of these cations: they just fit into the hollow between four F or O atoms of the non-bridging octahedron vertices (Fig. 90). Larger cations such as Cs^+ or Ba^{2+} fit if the octahedra are widened because of large central atoms, as for example in Cs_2UO_4 or Ba_2PbO_4. The composition A_2MX_4 is fulfilled when all the hollows between the octahedron apexes on either side of the $[MX_4]^{2n-}$ layer are occupied with A^{n+} ions. In the stacking of this kind of layer, the A^{n+} ions of one layer

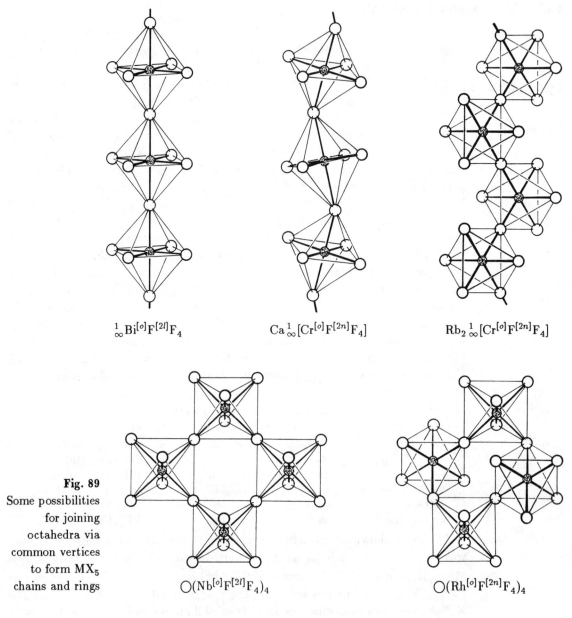

$\frac{1}{\infty}\mathrm{Bi}^{[o]}\mathrm{F}^{[2l]}\mathrm{F}_4$ $\mathrm{Ca}\frac{1}{\infty}[\mathrm{Cr}^{[o]}\mathrm{F}^{[2n]}\mathrm{F}_4]$ $\mathrm{Rb}_2\frac{1}{\infty}[\mathrm{Cr}^{[o]}\mathrm{F}^{[2n]}\mathrm{F}_4]$

Fig. 89
Some possibilities
for joining
octahedra via
common vertices
to form MX_5
chains and rings

$\bigcirc(\mathrm{Nb}^{[o]}\mathrm{F}^{[2l]}\mathrm{F}_4)_4$ $\bigcirc(\mathrm{Rh}^{[o]}\mathrm{F}^{[2n]}\mathrm{F}_4)_4$

are placed exactly above the X atoms of the preceding layer. Every A^{n+} ion then has coordination number 9 (four of the bridging atoms in the layer, the four X atoms of the surrounding octahedron apexes and the one X atom of the next layer); the coordination polyhedron is a capped square antiprism.

If we stack MX_4 layers with octahedron apex on top of octahedron apex and amalgamate the apexes with each other, the result is the three-dimensional network of the ReO_3 structure (Fig. 91). In this structure every octahedron shares all of its vertices with other octahedra; the bond angles at the bridging

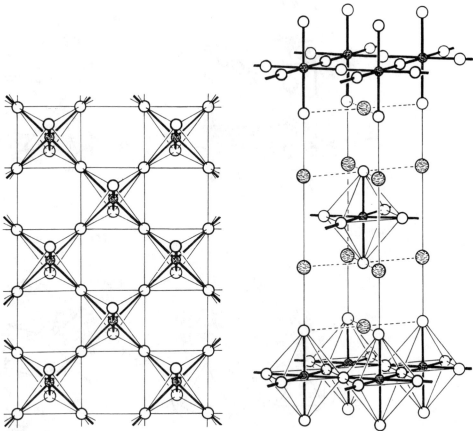

Fig. 90

MX_4 layer of vertex-sharing octahedra, and the packing of such layers in the K_2NiF_4 type. The packing in SnF_4 is obtained by leaving out the K^+ ions and shifting the layers towards each other in such a way that every octahedron apex of one layer comes to be between four apexes of the next layer

atoms are 180°. The centers of eight octahedra form a cube which corresponds to the unit cell. There is a rather large cavity in the center of the unit cell. This cavity can be occupied by a cation, which gives the perovskite type (perovskite = $CaTiO_3$); this structure type is rather frequent among compounds of the compositions AMF_3 and AMO_3, and because of its importance we discuss it separately (section 16.4, p. 197).

The degree of space filling of the ReO_3 type can be increased by rotating the octahedra about the direction of one of the space diagonals of the cubic unit cell (Fig. 91). Thereby the large cavity in the ReO_3 cell becomes smaller, the octahedra come closer to each other, and the bond angles at the bridging atoms decrease from 180° to 132°. Once this value is reached, we have the RhF_3 type, in which the F atoms are arranged as in a hexagonal closest-packing. A number of trifluorides crystallize with structures between these two extreme cases, with bond angles of about 150° at the F atoms: GaF_3, TiF_3, VF_3, CrF_3, FeF_3, CoF_3,

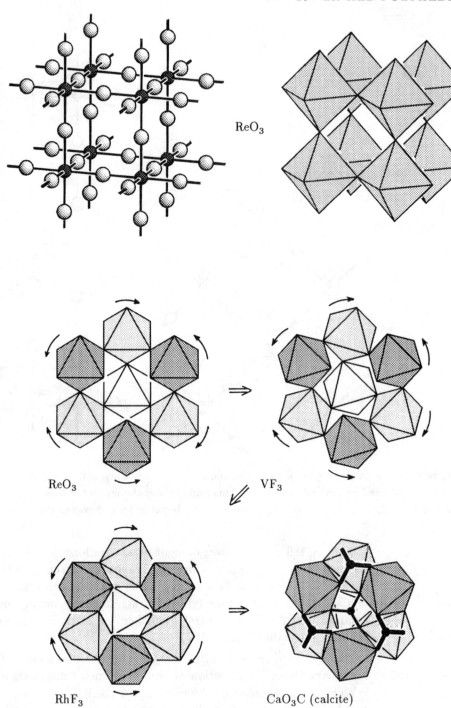

ReO$_3$

ReO$_3$ $\Rightarrow$ VF$_3$

RhF$_3$ $\Rightarrow$ CaO$_3$C (calcite)

Fig. 91

Top row: framework of octahedra sharing all vertices = ReO$_3$ type. Center and bottom: by rotating the octahedra, the ReO$_3$ type is converted to the VF$_3$, RhF$_3$ and calcite type and others. Some of them, such as ScF_3, are near to the ReO_3 type; others, such as MoF_3, are nearer to the hexagonal closest-packed structure. The mutual rotation of the octahedra can be continued to a "super dense" sphere packing which contains groups which have three squeezed atoms each. This corresponds to the structure of calcite ($CaCO_3$); the carbon atom of the carbonate ion is located in the center between three squeezed O atoms.

The $LiSbF_6$ type results when the sites of the metal atoms in the VF_3 type are occupied alternately by atoms of two different elements; this structure type is frequent among compounds AMF_6, e.g. $ZnSnF_6$. Similarly, two kinds of metal atoms can alternate in the metal positions of RhF_3 packing; this applies to PdF_3 (and PtF_3), which has to be regarded as $Pd^{II}Pd^{IV}F_6$ or $Pd^{2+}[PdF_6]^{2-}$, as can be seen by the different Pd–F bond distances of 217 pm (Pd^{II}) and 190 pm (Pd^{IV}).*

WO_3 occurs in a greater variety of modifications, all of which are distorted forms of the ReO_3 type (with W atoms shifted from the octahedron centers and with varying W–O bond lengths). In addition, a form exists that can be obtained by dehydrating $WO_3 \cdot \frac{1}{3}H_2O$; its framework is shown in Fig. 92. This also consists of vertex-sharing octahedra, with W–O–W bond angles of 150°. This structure is remarkable because of the channels it contains. These channels can be occupied by potassium ions in varying amounts, resulting in compositions K_xWO_3 ranging from $x = 0$ to $x = 0.33$ (rubidium and caesium ions can also be included). These compounds are termed hexagonal *tungsten bronzes*. Cubic tungsten bronzes have the ReO_3 structure with partial occupation of the voids by Li^+ or Na^+, i.e. they are intermediate between the ReO_3 type and

Fig. 92

Linking of the octahedra in hexagonal and tetragonal tungsten bronzes M_xWO_3. In the direction of view, the octahedra are arranged one on top of the other with common vertices. The channels in this direction contain varying amounts of alkali ions

*For literature concerning fluoride structures see [124].

the perovskite type. Tetragonal tungsten bronzes are similar to the hexagonal bronzes, but have narrower four and five-sided channels that can take up Na^+ or K^+ (Fig. 92). Tungsten bronzes are metallic conductors, and have metallic luster and colors that go from gold to black, depending on composition. They are very resistant chemically and serve as industrial catalysts and as pigments in "bronze colors".

15.2   Edge-sharing Octahedra

Two octahedra sharing one edge correspond to the composition $(MX_5)_2$ (or $(MX_{4/1}X_{2/2})_2$). This is the kind of structure common among pentahalides and ions $[MX_5]_2^{n-}$ when X = Cl, Br or I:

$(NbCl_5)_2$ $(TaCl_5)_2$ $(MoCl_5)_2$ $(WCl_5)_2$ $(ReCl_5)_2$ $(OsCl_5)_2$ $(UCl_5)_2$

$(NbBr_5)_2$ $(TaBr_5)_2$ $(UBr_5)_2$

$(NbI_5)_2$ $(TaI_5)_2$ $(PaI_5)_2$

$[TiCl_5]_2^{2-}$ $[ZrCl_5]_2^{2-}$ $[MoCl_5]_2^{2-}$

There do exist some exceptions in which the metal atoms are not coordinated octahedrally: $SbCl_5$ (monomer), PCl_5 (ionic $PCl_4^+PCl_6^-$), and PBr_5 (ionic $PBr_4^+Br^-$). $(MX_5)_2$ molecules can be packed very efficiently in such a way that the X atoms for themselves form a closest packing of spheres.

If the linking is continued to form a string of edge-sharing octahedra, the resulting composition is $MX_{2/1}X_{4/2}$, i.e. MX_4. Every octahedron then has two common edges with other octahedra in addition to two terminal X atoms. If the two terminal X atoms have a mutual *trans* arrangement, the chain is linear (Fig. 93). This kind of a chain occurs among tetrachlorides and tetraiodides if metal–metal bonds form pairwise between the M atoms of adjacent octahedra; the metal atoms then are shifted from the octahedron centers in the direction of the corresponding octahedron edge, and the octahedra experience some distortion.

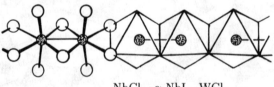

NbCl$_4$, α-NbI$_4$, WCl$_4$

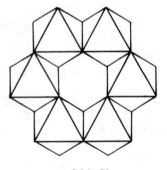

β-MoCl$_4$

Fig. 93
Some
configurations of
chains with
composition
MX_4 consisting
of edge-sharing
octahedra

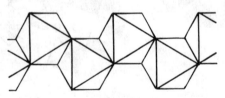

ZrCl$_4$, PtCl$_4$, UI$_4$ and others

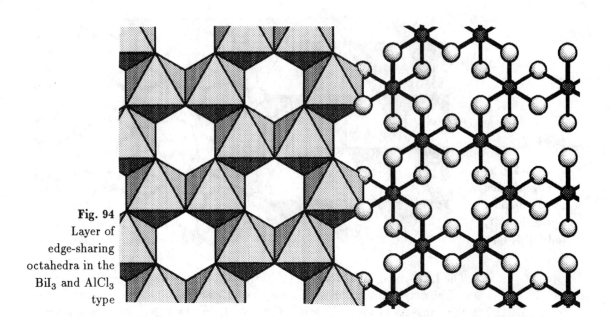

Fig. 94
Layer of
edge-sharing
octahedra in the
BiI$_3$ and AlCl$_3$
type

Examples are NbCl$_4$, NbI$_4$, WCl$_4$. The same kind of chain, but with metal atoms in the octahedron centers, has been observed for OsCl$_4$.

If the two terminal X atoms of an octahedron in an MX$_4$ chain have a *cis* arrangement, the chain can have a large variety of configurations. The most frequent one is a zigzag chain (Fig. 93); known examples include ZrCl$_4$, TcCl$_4$, PtCl$_4$, PtI$_4$, and UI$_4$. Chains with other configurations as in ZrI$_4$ are rare. Six edge-sharing octahedra can also join to form a ring (Fig. 93), but this kind of a structure is only known for one modification of MoCl$_4$.

By linking edge-sharing octahedra to form a layer as in Fig. 94, all X atoms act as bridging agents, and every one of them simultaneously belongs to two octahedra. This kind of a layer encloses voids that have octahedral shape (see also the figure on the cover). The composition of the layer is MX$_3$ (MX$_{6/2}$). Numerous trichlorides, tribromides and triiodides as well as some trihydroxides are composed of layers of this kind. The layers are stacked in such a way that the X atoms, for themselves, form a closest sphere packing, namely:

BiI$_3$ type: hexagonal closest-packing of X atoms
     FeCl$_3$, CrBr$_3$, Al(OH)$_3$ (bayerite) and others

AlCl$_3$ type: cubic closest-packing of X atoms
     YCl$_3$, CrCl$_3$ (high temp.) and others

The same kind of layers also occur in a second modification of Al(OH)$_3$, hydrargillite (gibbsite), but with a stacking in which adjacent O atoms of two layers are exactly one on top of the other; they are probably joined via hydrogen bridges.

Fig. 95
Edge-sharing
octahedra in a
layer of the CdI_2
and $CdCl_2$ type

The $CdCl_2$ and the CdI_2 type ($Cd(OH)_2$ type*) are also layer structures (Fig. 95). The octahedra in the layer share six edges each. The structure of the layer is the same as in an MX_3 layer if the voids in the MX_3 layer were occupied by M atoms. Every halogen atom is shared by three octahedra ($MX_{6/3}$). The stacking of the layers is also of the same kind:

CdI_2 type ($Cd(OH)_2$ type): hexagonal closest-packing of X atoms

$MgBr_2$, $TiBr_2$, VBr_2, $CrBr_2^\dagger$, $MnBr_2$, $FeBr_2$, $CoBr_2$, $NiBr_2$, $CuBr_2^\dagger$
MgI_2, CaI_2, PbI_2, TiI_2, VI_2, $CrI_2^\dagger$, MnI_2, FeI_2, CoI_2
$Mg(OH)_2$, $Ca(OH)_2$, $Mn(OH)_2$, $Fe(OH)_2$, $Co(OH)_2$, $Ni(OH)_2$, $Cd(OH)_2$
SnS_2, TiS_2, ZrS_2, NbS_2, PtS_2
$TiSe_2$, $ZrSe_2$, $PtSe_2$
Ag_2F, Ag_2O (F and O in the octahedron centers, respectively)

$CdCl_2$ type: cubic closest-packing of X atoms

$MgCl_2$, $MnCl_2$, $FeCl_2$, $CoCl_2$, $NiCl_2$
Cs_2O (O in the octahedron centers)

$\dagger$ distorted by Jahn-Teller effect

Among the hydroxides such as $Mg(OH)_2$ (brucite) and $Ca(OH)_2$ the packing of the O atoms deviates from an ideal hexagonal closest-packing in that the layers are somewhat flattened; the bond angles M–O–M in the layer are larger than the ideal 90° for undistorted octahedra (e.g. 98.5° in $Ca(OH)_2$).

15.3 Face-sharing Octahedra

Two octahedra sharing a common face correspond to a composition M_2X_9 (Fig. 96). This structure is known for some molecules, for example $Fe_2(CO)_9$, and especially for some ions with trivalent metals. In some cases, the reason for

---

*Because there exist numerous stacking variants of CdI_2, some authors prefer the term $Cd(OH)_2$ type.

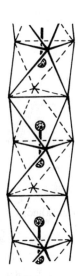

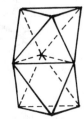

Fig. 96

Two face-sharing octahedra in ions $[M_2X_9]^{3-}$
and a string of face-sharing octahedra in ZrI_3

the face-sharing is obviously the presence of metal–metal bonds, for example in the $[W_2Cl_9]^{3-}$ ion; its small magnetic moment suggests a W≡W bond (cf. the structure shown on p. 158). The $[Cr_2Cl_9]^{3-}$ ion has the same structure, but nevertheless, it exhibits the paramagnetism that is to be expected for the electron configuration d^3; $[Mo_2Cl_9]^{3-}$ is intermediate in its behaviour. The bond angles at the bridging atoms also reflect these differences: 58° in $[W_2Cl_9]^{3-}$, 77° in $[Cr_2Cl_9]^{3-}$. $[M_2X_9]^{3-}$ ions are also known among compounds which have no metal–metal bonding, for example $[Tl_2Cl_9]^{3-}$ or $[Bi_2Br_9]^{3-}$. The occurrence of this kind of ion often depends on the counter-ion, i.e. the packing in the crystal is an important factor. For example, Cs^+ ions and Cl^- ions, being of comparable size, allow for a close packing and facilitate the occurrence of these double octahedra. They also occur with large cations such as $P(C_6H_5)_4^+$.

If opposite faces of the octahedra are used to continue their linking, the result is a strand of composition MX_3 (Fig. 96). Strands of this kind occur among some trihalides, when metal–metal bonds between pairs of adjacent octahedra are present: β-$TiCl_3$, TiI_3, ZrI_3, $MoBr_3$, $RuBr_3$. Anionic strands of the same kind are also known in compounds such as $Cs[NiCl_3]$ or $Ba[NiO_3]$; again, the comparable sizes of the cations Cs^+ and Ba^{2+} and the anions Cl^- and O^{2-}, respectively, facilitate a close packing.

15.4   Octahedra Sharing Vertices and Edges

Units $(MX_5)_2$ can be joined via common vertices to form double strands as shown in Fig. 97. Since every octahedron still has two terminal atoms, the composition $MX_{2/1}X_{2/2}Z_{2/2}$ or MX_3Z results, the atoms in the common vertices being designated by Z. The same structure also results when two parallel strands of the BiF_5 type are joined via common edges. Compounds such as $NbOCl_3$ or WOI_3 have this kind of structure, with oxygen atoms taking the positions of the common vertices.

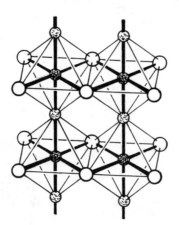

Fig. 97
Linking of octahedra in NbOCl$_3$

In rutile every O atom is common to three octahedra, as expressed by the formula TiO$_{6/3}$. As can be seen in Fig. 98, linear strands of edge-sharing octahedra are present, as in compounds MX$_4$. Parallel strands are joined by common octahedron vertices. Compared with the layers of the CdI$_2$ type, the number of common edges is reduced, namely two instead of six per octahedron. According to PAULING's third rule (p. 45), this favors the rutile type for electrostatic reasons. Compounds MX$_2$ with octahedrally coordinated M atoms therefore prefer the rutile type to the CdCl$_2$ or CdI$_2$ type if they are very polar: among dioxides and difluorides the rutile type is very common, examples being GeO$_2$, SnO$_2$, CrO$_2$, MnO$_2$, and RuO$_2$ as well as MgF$_2$, FeF$_2$, CoF$_2$, NiF$_2$, and ZnF$_2$.

In the normal rutile type the metal atoms in a strand of edge-sharing octahedra are equidistant. In some dioxides, however, alternating short and long M–M distances occur, i. e. the metal atoms are pairwise shifted from the octahedron centers towards each other. This phenomenon occurs (though not always) when the metal atoms still have d electrons and thus can engage in metal–metal bonds, for example in the low-temperature modifications of VO$_2$, NbO$_2$, MoO$_2$, and WO$_2$ (the high-temperature forms have the normal rutile structure).

Fig. 98
Strands of edge-sharing octahedra are joined with each other in rutile and α-PbO$_2$ via common vertices (marked by dots)

rutile

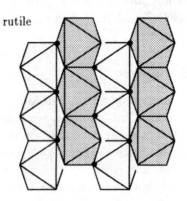

α-PbO$_2$

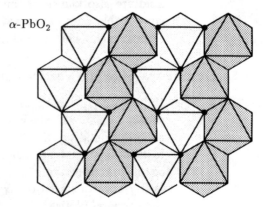

The zigzag chains of edge-sharing octahedra that occur among compounds MX_4 can also be joined by common vertices, resulting in the α-PbO_2 type (Fig. 98); this structure type is less frequent.

Octahedra linked in different ways often occur when different kinds of metal atom are present. Li_2ZrF_6 offers an example. The Li and the F atoms are arranged in layers of the same kind as in BiI_3; the layers are joined by single ZrF_6 octahedra which are placed below and on top of the voids of the layer (Fig. 99). The octahedra of the Li_2F_6 layer share common edges with one another and they share vertices with the ZrF_6 octahedra.

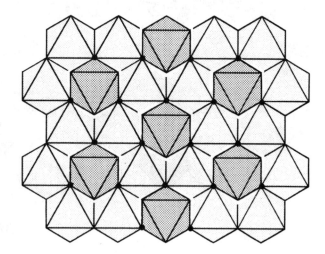

Fig. 99
Linking of octahedra in Li_2ZrF_6 and Sn_2PbO_6

Isopoly and Heteropoly Acids

Numerous linking patterns, some of which are very complex, consisting mainly of octahedra sharing vertices and edges are known among polyvanadates, niobates, tantalates, molybdates, and tungstates. If only one of these elements occurs in the polyhedra, they are also called isopoly acids or isopoly anions. If additional elements also form part of the structures, they are called heteropoly acids; the additional atoms can be coordinated tetrahedrally, octahedrally, square-antiprismatically or icosahedrally. The dodecamolybdatophosphate $[PO_4Mo_{12}O_{36}]^{3-}$ is the classical example of this compound class; the precipitation of its ammonium salt serves as an analytical proof for phosphate ions. It has the KEGGIN structure: four groups consisting of three edge-sharing octahedra are linked by common vertices, so that the twelve MoO_6 octahedra form a cage (Fig. 100). The tetrahedrally coordinated phosphorus atom is in the interior of the cage; it can be replaced by Al(III), Si(IV), As(V), Fe(III), and others. Octahedrally coordinated hetero atoms are found in the ions $[EMo_6O_{24}]^{n-}$, e.g. $[TeMo_6O_{24}]^{5-}$ (E = Te(VI), I(VII), Mn(IV); Fig. 100).

Some isopoly anions consist of a compact system of edge-sharing octahedra; a few examples are shown in Fig. 100. Oxygen atoms with high coordination

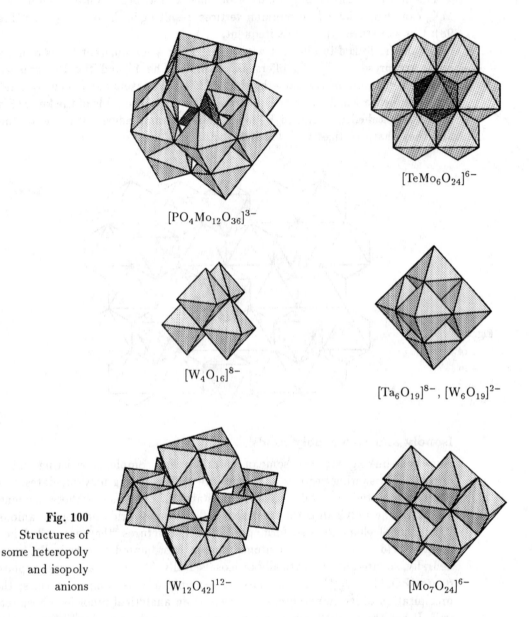

$[PO_4Mo_{12}O_{36}]^{3-}$

$[TeMo_6O_{24}]^{6-}$

$[W_4O_{16}]^{8-}$

$[Ta_6O_{19}]^{8-}$, $[W_6O_{19}]^{2-}$

Fig. 100
Structures of
some heteropoly
and isopoly
anions

$[W_{12}O_{42}]^{12-}$

$[Mo_7O_{24}]^{6-}$

numbers are situated in their interior; for example, the O atom at the center
of the $[W_6O_{19}]^{2-}$ ion has coordination number 6. Other representatives with
part of their octahedra sharing only vertices can have more or less large cavities
in their interior, such as, for example, the $[W_{12}O_{42}]^{2-}$ ion. Isopoly anions are
formed in aqueous solutions, depending on the pH value. Molybdate solutions,
for example, contain MoO_4^{2-} ions at high pH values, $[Mo_7O_{24}]^{6-}$ ions at pH $\approx$ 5
and even larger aggregates in more acid solutions.

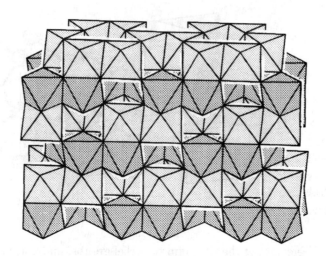

Fig. 101
Linked octahedra
in corundum
(α-Al$_2$O$_3$) and in
ilmenite
(FeTiO$_3$; Fe
octahedra light,
Ti octahedra
dark)

15.5 Octahedra Sharing Edges and Faces

The corundum structure (α-Al$_2$O$_3$) is the result of linking layers of the BiI$_3$ kind one on top of the other; the layers are mutually shifted as shown in Fig. 101. There are pairs of face-sharing octahedra, and, in addition, every octahedron shares three edges within a layer and three vertices with octahedra from the adjacent layer to which it has no face-sharing connection. This structure type is adopted by some oxides M$_2$O$_3$ (e.g. Ti$_2$O$_3$, Cr$_2$O$_3$, α-Fe$_2$O$_3$).

Alternate layers can be occupied by two different kinds of metal atom, then every pair of the face-sharing octahedra contains two different metal atoms; this is the ilmenite type (FeTiO$_3$). Ilmenite is, along with perovskite, another structure type for the composition A^{II}M^{IV}O$_3$. The space for the A^{2+} ion is larger in perovskite. Which structure type is preferred can be estimated with the aid of the ionic radius ratio:

$$r(A^{2+})/r(O^{2-}) < 0.7 \qquad \text{ilmenite}$$
$$r(A^{2+})/r(O^{2-}) > 0.7 \qquad \text{perovskite}$$

Another criterion for the same purpose is discussed on p. 200.

The **nickel arsenide type** (NiAs) is the result of linking layers of the kind as in cadmium iodide. Continuous strands of face-sharing octahedra perpendicular to the layers arise from this connection. The same structure results when strands of the kind found in ZrI$_3$ are linked from all sides by common edges (Fig. 102). The nickel atoms in the octahedron centers form a primitive hexagonal lattice; every arsenic atom is surrounded by a trigonal prism of Ni atoms. Since the atoms in the face-sharing octahedra are quite close to one another, interactions must exist between them. This shows up in the electric properties: compounds with the NiAs structure are semiconductors or metallic conductors. Numerous representatives are known for this structure type: the metallic component is an

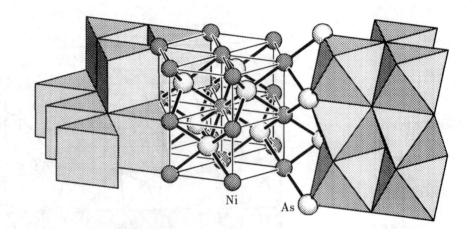

Fig. 102
Octahedra and
trigonal prisms
in the NiAs
structure

element of the titanium to nickel groups, and Ga, Si, P, S, and their heavier homologous elements are adequate to substitute for the arsenic. Examples include TiS, TiP, CoS, and CrSb.

15.6   Linked Trigonal Prisms

In NiAs the Ni atoms form a network of trigonal prisms which contain the As atoms, and the Ni atoms have octahedral coordination. Metal atoms with trigonal-prismatic coordination are present in MoS_2. The S atoms form hexagonal planes with the stacking sequence $AABBAABB\ldots$ or $AABBCC\ldots$ (or some additional stacking variants). In every pair of congruent planes, e.g. AA, there are edge-sharing trigonal prisms which contain the Mo atoms and form a layer (Fig. 103). Between the layers, i.e. between sulfur atom planes having different positions, e.g. AB, the only attractive forces are weak VAN DER WAALS interactions. The MoS_2 layers can easily slip as in graphite; for this reason MoS_2 is used as a lubricant. Further similarities to graphite include the anisotropic electrical conductivity and the ability to form intercalation compounds, e.g. $K_{0.5}MoS_2$.

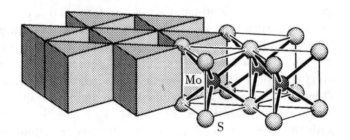

Fig. 103
Layer of
edge-sharing
trigonal prisms
in MoS_2

15.7   Vertex-sharing Tetrahedra. Silicates

The linking of tetrahedra takes place predominantly by sharing vertices. Edge-sharing and especially face-sharing is considerably less frequent than among octahedra.

Two tetrahedra sharing a common vertex form a unit M_2X_7. This unit is known among oxides such as Cl_2O_7 and Mn_2O_7 and among several anions, e.g. $S_2O_7^{2-}$, $Cr_2O_7^{2-}$, $P_2O_7^{4-}$, $Si_2O_7^{6-}$, and $Al_2Cl_7^-$. Depending on the conformation of the two tetrahedra, the bond angle at the bridging atom can have values between 102.1° and 180° (Fig. 104).

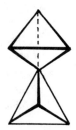
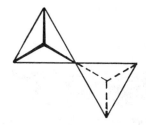

Fig. 104 Different conformations of two vertex-sharing tetrahedra

A chain of vertex-sharing tetrahedra results when every tetrahedron has two terminal and two bridging atoms; the composition is $MX_{2/1}X_{2/2}$ or MX_3. The chain can be closed to form a ring as in $[SO_3]_3$, $[PO_3^-]_3$, $[SiO_3^{2-}]_3$ or $[SiO_3^{2-}]_6$. Endless chains have different shapes depending on the mutual conformation of the tetrahedra (Fig. 105). They occur especially among silicates, where the chain shape is also determined by the interactions with the cations. In silicates of the composition $MSiO_3$ with octahedrally coordinated M^{2+} ions (M^{2+} = Mg^{2+}, Ca^{2+}, Fe^{2+}, and others, ionic radii 50 to 100 pm), the co-ordination octahedra of the metal ions are arranged to form layers as in $Mg(OH)_2$. Thus the octahedra share edges, and their vertices are also shared with vertices of the SiO_3^{2-} chains corresponding to terminal O atoms of the chain; these O atoms thus link tetrahedra with octahedra. Depending on the kind of cation, i.e. octahedron size, different chain conformations occur (Fig. 105). Compounds of this kind such as enstatite, $MgSiO_3$, are termed pyroxenes if the silicate chain is a *Zweierkette*, i.e. if the chain pattern repeats after two tetrahedra; pyroxenoids have more complicated chain forms, for example the *Dreierkette* in wollastonite, $CaSiO_3$.*

Tetrahedra linked via three vertices correspond to a composition $MX_{1/1}X_{3/2}$ or $MX_{2,5}$ = M_2X_5. Small units consisting of four tetrahedra are known in P_4O_{10}, but most important are the layer structures in the numerous sheet silicates and alu-

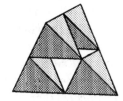

Linked tetrahedra in P_4O_{10}

---

*From the German *zwei* = two, *drei* = three, *zweier* = two-membered, *dreier* = three-membered, *Kette* = chain.

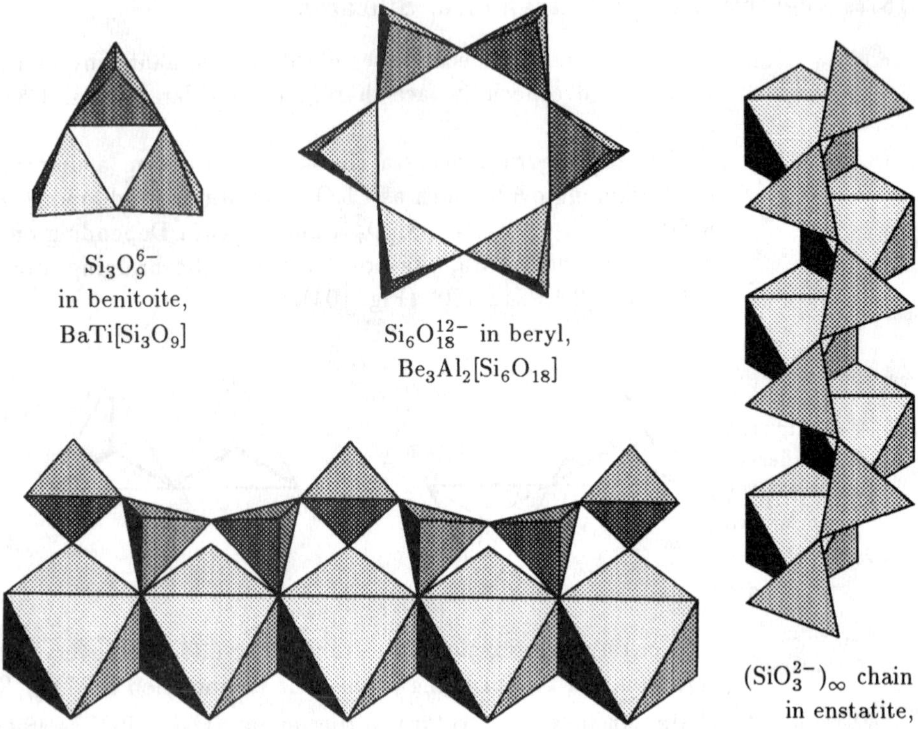

Fig. 105
Some forms of rings and chains of vertex-sharing tetrahedra in silicates. How the chain conformations adapt to the size of the cation octahedra is shown for two chains (the octahedron chain is a section of a layer)

$Si_3O_9^{6-}$ in benitoite, $BaTi[Si_3O_9]$

$Si_6O_{18}^{12-}$ in beryl, $Be_3Al_2[Si_6O_{18}]$

$(SiO_3^{2-})_\infty$ in wollastonite, $CaSiO_3$

$(SiO_3^{2-})_\infty$ chain in enstatite, $MgSiO_3$

minosilicates with anions of the composition $[Si_2O_5^{2-}]_\infty$ and $[AlSiO_5^{3-}]_\infty$, respectively. Because the terminal vertices of the single tetrahedra can be oriented in different sequences to one or the other side of the layer, a large number of structural varieties are possible; moreover, the layers can be corrugated (Fig. 106).

Sheet silicates are of frequent natural occurrence, the most important ones being clay minerals (prototype: kaolinite), talc (soapstone) and micas (prototype: muscovite). In these minerals the terminal O atoms of a silicate layer are bonded with octahedrally coordinated cations; these are mainly Mg^{2+}, Ca^{2+}, Al^{3+}, or Fe^{2+}. The octahedra are linked with each other by common edges, forming layers as in $Mg(OH)_2$ ($\cong CdI_2$) or $Al(OH)_3$ ($\cong BiI_3$). The number of terminal O atoms in the silicate layer is not sufficient to provide all of the O atoms for the octahedron layer, so the remaining octahedron vertices are occupied by additional OH^- ions. Two kinds of linking between the silicate layer and the octahedron layer can be distinguished: in *cation-rich sheet silicates* an octahedron layer is linked with only one silicate layer on one of its sides, and the result is an octahedron-tetrahedron sheet; in *cation-poor sheet silicates* there are tetrahedron-octahedron-tetrahedron sheets, an octahedron layer being linked to a silicate layer on both sides (Fig. 107). Depending on whether the cation layer is of the $Mg(OH)_2$ or the $Al(OH)_3$ type, certain numerical relations result be-

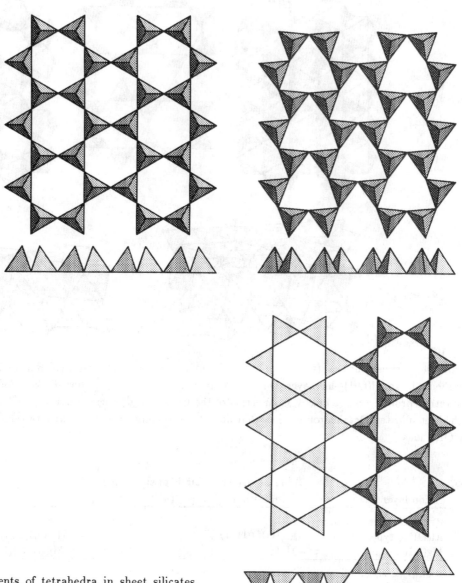

Fig. 106
Some arrangements of tetrahedra in sheet silicates.
The lower image in each case represents a side view

tween the number of cations in the octahedron layer and the number of silicate tetrahedra; the number of OH^- ions needed to complete the octahedra is also fixed by geometry. Additional cations may be intercalated between the sheets (Table 21).

The sheets consisting of tetrahedron-octahedron-tetrahedron layers in cation poor-sheet silicates are completely planar due to the symmetrical environment of the cation layer. If the sheets are electrically neutral as in talc, the attractive

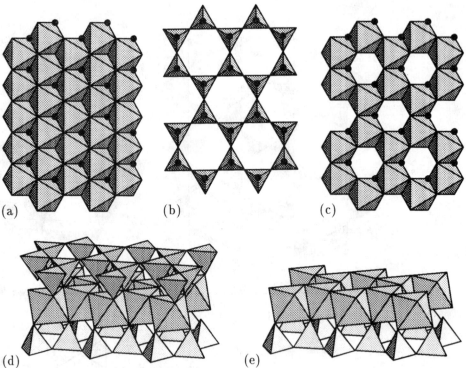

Fig. 107

Structures of sheet silicates. (**a**) Cation octahedra in an $Mg(OH)_2$-like layer. (**b**) Silicate layer. (**c**) Cation octahedra in an $Al(OH)_3$-like layer. Dots mark the polyhedron vertices shared when the octahedron and the tetrahedron layers are linked. Connection of the layers in: (**d**) cation-poor sheet silicates, (**e**) cation-rich sheet silicates. Octahedron vertices that do not act as common vertices with tetrahedra are occupied by OH^- ions

Table 21 Different kinds of sheet silicates

cation layer	composition	examples
cation-rich sheet silicates		
$Al(OH)_3$ type	$M_2(OH)_4[T_2O_5]$	kaolinite, $Al_2(OH)_4[Si_2O_5]$
$Mg(OH)_2$ type	$M_3(OH)_4[T_2O_5]$	chrysotile, $Mg_3(OH)_4[Si_2O_5]$
cation-poor sheet silicates		
$Al(OH)_3$ type	$M_2(OH)_2[T_4O_{10}]$	pyrophyllite, $Al_2(OH)_2[Si_4O_{10}]$
$Mg(OH)_2$ type	$M_3(OH)_2[T_4O_{10}]$	talc (soapstone), $Mg_3(OH)_2[Si_4O_{10}]$
cation-poor sheet silicates with intercalations		
$Al(OH)_3$ type	$A\{M_2(OH)_2[T_4O_{10}]\}$	muscovite, $K\{Al_2(OH)_2[AlSi_3O_{10}]\}$
	$A_x\{M_2(OH)_2[T_4O_{10}]\} \cdot nH_2O$	montmorillonite,
		$Na_x\{Mg_xAl_{2-x}(OH)_2[Si_4O_{10}]\} \cdot nH_2O$
$Mg(OH)_2$ type	$M_3(OH)_2[T_4O_{10}] \cdot nH_2O$	vermiculite, $Mg_3(OH)_2[Si_4O_{10}] \cdot nH_2O$

T = tetrahedrally coordinated Al or Si
M = Mg^{2+}, Ca^{2+}, Al^{3+}, Fe^{2+} etc.
A = Na^+, K^+, Ca^{2+} etc.

forces between them are weak; as a consequence, the crystals are soft and easy to cleave. The use of talc as powder, lubricating agent, polishing material and filling material for paper is due to these properties.

Micas are cation-poor sheet silicates consisting of electrically charged sheets that are being held together by intercalated, unhydrated cations. For this reason the sheets cannot slip as in talc, but they still can be cleaved sheetwise. The crystals usually form thin, stiff plates. Larger plates (in sizes from centimeters to meters) are used industrially because of their ruggedness, transparency, electrical insulating properties, and chemical and thermal resistance (muscovite up to approximately 500 °C, phlogopite, $KMg_3(OH)_2[AlSi_3O_{10}]$, up to approximately 1000 °C).

Clay materials show a different behaviour; they are either cation-poor or cation-rich sheet silicates. They have the ability to swell by taking up varying amounts of water in between the sheets. If they contain intercalated, hydrated cations as in montmorillonite, they act as cation exchangers. Montmorillonite, especially when it has intercalated Ca^{2+} ions, has thixotropic properties and is used to seal up drill holes. The effect is due to the charge distribution on the crystal platelets: they bear a negative charge on the surface and a positive charge on the edges. In suspension they therefore orient themselves edge against surface, resulting in a jelly. Upon agitation the mutual orientation is disturbed and the mass is liquefied.

Swollen clay materials are soft and easy to mould. They serve to produce ceramic materials. High quality fire-clay has a high kaolinite content. Upon firing, the intercalated water is removed first at approximately 100 °C. Then, beginning at 450 °C, the OH groups are converted to oxidic O atoms by liberation of water, and after some more intermediate steps, mullite is formed at approximately 950 °C. Mullite is an aluminum aluminosilicate, $Al_{(4-x)/3}[Al_{2-x}Si_xO_5]$ with $x \approx 0.6$ to 0.8.

Because the dimensions of an octahedron and a tetrahedron layer usually do not coincide exactly, the unilateral linking of the layers in cation-rich sheet silicates leads to tensions. As long as the dimensions do not deviate too much, the tension are relieved by slight rotations of the tetrahedra and the sheets remain planar. This applies to kaolinite, which has only Al^{3+} ions in the cation layer. With the larger Mg^{2+} ions the metric fit is inferior; the tension then causes a bending of the sheets. This can be compensated for by tetrahedron apexes periodically pointing to one side and then to the other side of the tetrahedron layer, as in antigorite (Fig. 108). If the bending is not compensated for, the sheets curl up to form tubes in the way shown in Fig. 108, corresponding to the structure of chrysotile, $Mg_3(OH)_4[Si_2O_5]$. Because the sheet only tolerates curvatures within certain limits and the curvature is less on the inside than on the outside of the tube, the tubes remain hollow and they cannot exceed some maximum diameter. The inner diameter in chrysotile is about 5 nm, the outer one 20 nm. The tubular building blocks explain the fibrous properties of chrysotile which used to be the most important asbestos mineral.

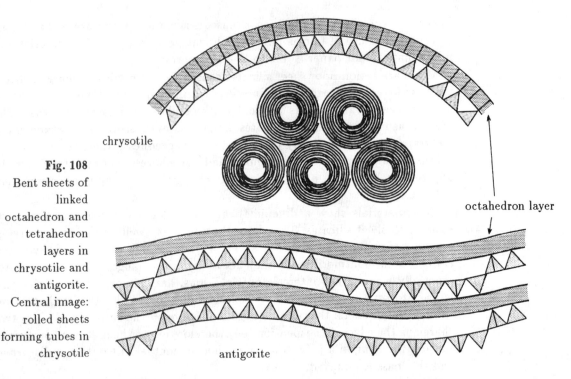

chrysotile

Fig. 108
Bent sheets of
linked
octahedron and
tetrahedron
layers in
chrysotile and
antigorite.
Central image:
rolled sheets
forming tubes in
chrysotile

octahedron layer

antigorite

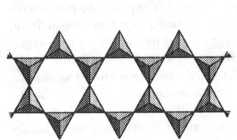

$[Si_4O_{11}^{6-}]_n$ ribbon

A layer of tetrahedra can be considered as being built up by linking parallel chains. That this is not a mere formalism is shown by the existence of intermediate stages. Two linked silicate chains result in a ribbon of the composition $[Si_4O_{11}^{6-}]_n$; it has two kinds of tetrahedra, one kind being joined via three and the other kind via two vertices, $[SiO_{1/1}O_{3/2}SiO_{2/1}O_{2/2}]^{3-}$. Silicates of this type are termed amphiboles. They are fibrous and also used to be used as asbetsos.

Linking Tetrahedra by All Four Vertices. Zeolites

Mercury iodide offers an example of a layer structure consisting of tetrahedra sharing all of their vertices (Fig. 109). Much more frequent are framework structures; first of all, they include the different modifications of SiO_2 and the aluminosilicates that are discussed in section 11.5. Another important class of aluminosilicates are the zeolites. They occur as minerals, but are also produced industrially. They have structures consisting of certain polyhedra that are linked in such a way that hollows and channels of differing sizes and shapes are present.

Fig. 110 shows the structure of the methylpolysiloxane $Me_8Si_8O_{12}$ which can be made by hydrolysis of $MeSiCl_3$. Its framework is a cube of silicon atoms linked via oxygen atoms placed on every cube edge. The O atoms are situated slightly to one side of the edges, thus allowing for a framework without tensions

Fig. 109
Section of a layer
in HgI_2

and bond angles of 109.5° at the Si atoms and of 148.4° at the O atoms. This
framework has been drawn schematically in the rest of Fig. 110 as a simple
cube. It is one of several possible building units occurring in zeolites; the place
of the methyl groups is taken by O atoms which mediate the connection to
other Si atoms. In addition to the cube, other polyhedra occur, some of which
are shown in Fig. 110. Every vertex of these polyhedra is occupied by an Si
or Al atom, and in the middle of each edge there is an O atom which joins
two of the atoms in the vertices. In a zeolite four edges meet at every vertex,
corresponding to the four bonds of the tetrahedrally coordinated atoms.

The linking pattern of two zeolites is shown in Fig. 110. They have the "β-
cage" as one of their building blocks; that is a truncated octahedron, a polyhedron with 24 vertices and 14 faces. In the synthetic zeolite A (Linde A) the
β-cages form a cubic primitive lattice, and are joined by cubes. β-cages distributed in the same manner as the atoms in diamond and linked by hexagonal
prisms make up the structure of faujasite (zeolite X).

The fraction of aluminum atoms in the framework is variable. For every one
of them there is one negative charge. As a whole, the framework thus is a
polyanion; the cations occupy places in the hollows. This in principle applies
also to other aluminosilicates, but the framework of the zeolites is much more
open. This is the basis of the characteristic properties of the zeolites: they
act as cation exchangers and absorb and release water easily. A zeolite that
has been dehydrated by heating it in vacuo is highly hygroscopic and can be
used to remove water from solvents or gases. In addition to water, it can also
absorb other molecules; the size and shape of the molecules relative to the
size and shape of the hollow spaces in the zeolite determine how easily this
occurs and how tightly the guest molecules are retained by the host framework.
Different types of zeolite differ widely in regard to their hollows and channels,
and they can be made to measure in order to take up certain molecules. This
effect is applied for the selective separation of compounds, and therefore zeolites
are also termed molecular sieves. For example, they can separate unbranched
and branched alkanes, which is important for petroleum refineries. Even the

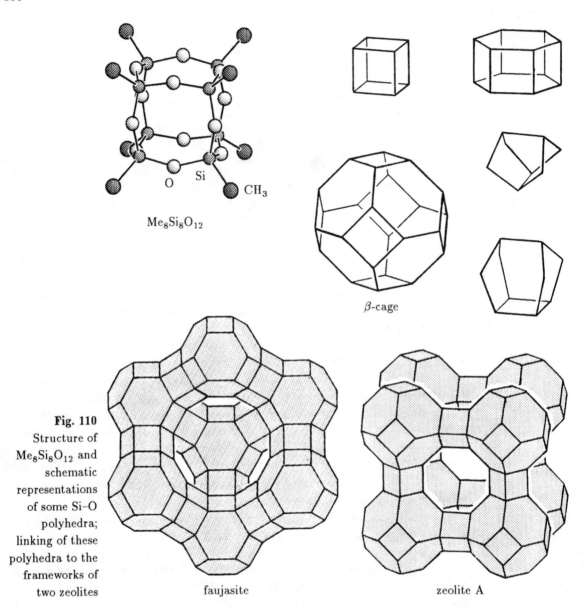

O   Si

CH₃

Me₈Si₈O₁₂

β-cage

Fig. 110
Structure of
Me₈Si₈O₁₂ and
schematic
representations
of some Si–O
polyhedra;
linking of these
polyhedra to the
frameworks of
two zeolites

faujasite   zeolite A

separation of O_2 and N_2 is possible. The channels can also accomodate different molecules simultaneously, the shape of the channels forcing the molecules to adopt some definite mutual orientation. As a consequence, zeolites can act as selective catalysts. Synthetic zeolite ZSM-5, for exmaple, serves to catalyze the hydrogenation of methanol to alkanes.

Zeolites are structurally related to colorless sodalite, $Na_4Cl[Al_3Si_3O_{12}]$, and to deeply colored ultramarines. These have aluminosilicate frameworks that enclose cations but no water molecules (Fig. 111). Their special feature is the additional presence of anions in the hollows, e.g. Cl^-, SO_4^{2-}, S_2^-, or S_3^-. The

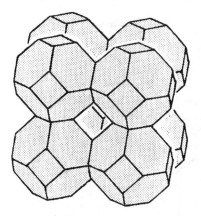

Fig. 111
Sodalite and ultramarine framework

two last-mentioned species are colored radical ions (green and blue, respectively) that are responsible for the brilliant colors. The best-known representative is the blue mineral lapis lazuli, $Na_4S_x[Al_3Si_3O_{12}]$, which is also produced industrially and serves as color pigment.

Framework silicates are also termed tectosilicates. Their common feature is the three-dimensional connection of tetrahedra sharing all four vertices. They are subdivided into:

1. Pyknolites, which have a framework with relatively small cavities that are filled with cations; for example: feldspars $M^+[AlSi_3O_8^-]$ and $M^{2+}[Al_2Si_2O_8^{2-}]$ such as $K[AlSi_3O_8]$ (orthoclase, sanidine) or plagioclase, $Ca_{1-x}Na_x[Al_{2-x}Si_{2+x}O_8]$ which includes $Na[AlSi_3O_8]$ (albite; $x = 1$) and $Ca[Al_2Si_2O_8]$ (anorthite; $x = 0$). Feldspars, especially plagioclase, are by far the most abundant minerals in Earth's crust.*

2. Clathrasils have polyhedral cavities, but with windows that are too small to allow the passage of other molecules, so that enclosed ions or foreign molecules cannot escape. Examples are ultramarines and melanophlogite $(SiO_2)_{46} \cdot 8\,(N_2, CO_2, CH_4)$.

3. Zeolites with polyhedral cavities which are connected by wide windows or channels that permit the diffusion of foreign ions or molecules.

The structural relationship between SiO_2 and H_2O (cf. section 11.5) also shows up in the clathrates (inclusion compounds); they include clathrasils that enclose foreign molecules. Water forms analogous clathrate hydrates which consist of foreign molecules enclosed by a framework of H_2O molecules. As in ice, every O atom is surrounded by four H atoms. The structures are only stable in the presence of the enclosed molecules, otherwise the hollow, wide-meshed framework would collapse. The gas hydrates are among the best-known species of this kind. They have particles such as Ar, CH_4, H_2S or Cl_2 enclosed by a framework which has two dodecahedral and six larger tetracaidecahedral (polyhedron

---

*The term feldspar expresses their occurrence; German *Feld* = field; spar (German *Spat*) is a mineralogical term for certain naturally occurring crystalline compounds.

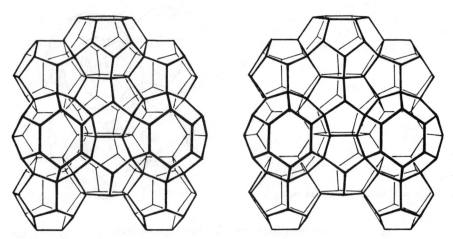

Fig. 112
Section of the framework in gas hydrates of type I. Every vertex represents an O atom, and along every edge there is an H atom (stereo image)

with 14 faces) cavities per 46 molecules of H_2O (Fig. 112; the above-mentioned melanophlogite has the same structure). If all of the cavities are occupied, the composition is $(H_2O)_{46}X_8$ or $X \cdot 5\frac{3}{4}H_2O$; if only the larger cavities are occupied, as in the Cl_2 hydrate, the composition is $(H_2O)_{46}(Cl_2)_2$ or $Cl_2 \cdot 7\frac{2}{3}H_2O$. Different frameworks with larger cavities form with larger foreign molecules. Examples include $(CH_3)_3CNH_2 \cdot 9\frac{3}{4}H_2O$, $HPF_6 \cdot 6H_2O$, and $CHCl_3 \cdot 17H_2O$. Clathrates such as $C_3H_8 \cdot 17H_2O$, which has a melting point of 8.5 °C, can crystallize from humid natural gas during cold weather and obstruct pipelines. The clathrate structure also occurs among the compounds Na_8Si_{46}, K_8Si_{46}, K_8Ge_{46}, and K_8Sn_{46}, the Si atoms taking the positions of the water molecules and thus having four bonds each. The alkali ions occupy the cavities, and their electrons contribute to a metallic electron gas.

15.8   Edge-sharing Tetrahedra

Two tetrahedra sharing one edge lead to the composition M_2X_6, as in Al_2Cl_6 (in the gaseous state or in solution) (Fig. 88, p. 158). Continuation of the linking using opposite edges results in a linear chain, with all X atoms having bridging functions. Chains of this kind are known in $BeCl_2$ and SiS_2 as well as in the anion of $K[FeS_2]$ (Fig. 113).

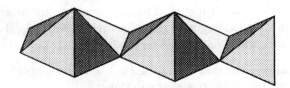

Fig. 113
Linked tetrahedra in SiS_2

If tetrahedra are joined via four of their edges, the resulting composition is $MX_{4/4}$ or MX. This kind of linking corresponds to the structure of the red modification of PbO, in which O atoms occupy the tetrahedron centers and the Pb atoms the vertices (Fig. 114). This rather peculiar structure may be regarded as a consequence of the steric influence of the lone electron pair at the Pb(II) atom; if we include the electron pair, the coordination polyhedron of a lead atom is a square pyramid. The layer can be described as a checker board having O atoms at its cross-points; the Pb atoms are placed above the white and under the black fields of the board.

The CaF_2 structure can be regarded as a network of three-dimensionally linked, edge-sharing FCa_4 tetrahedra (cf. Fig. 117 b).

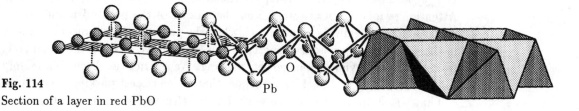

Fig. 114
Section of a layer in red PbO

15.9 Problems

15.1 Take $W_4O_{16}^{8-}$ ions (Fig. 100) and pile them to form a column consisting of pairs of edge-sharing octahedra that alternate crosswise. What is the composition of the resulting column?

15.2 Take pairs of face-sharing coordination octahedra and join them by common vertices to form a chain, with every octahedron taking part in one common vertex not belonging to the common face. What is the composition of the resulting chain?

15.3 What is the composition of a column of square antiprisms joined by common square faces?

15.4 Which of the following compounds could possibly form columns of face-sharing octahedra as in ZrI_3?
InF_3, $InCl_3$, MoF_3, MoI_3, TaS_3^{2-}

15.5 Take the network of vertex-sharing tetrahedra of the Cu atoms in $MgCu_2$ (Fig. 85) and assume that there is an additional atom inside of every tetrahedron. What structure type would this be?

16   Sphere Packings with Occupied Interstices

In chapter 15 the packing of polyhedral building blocks is mentioned repeatedly, for example in respect to the difference between the $CdCl_2$ and CdI_2 type. These both consist of the same kind of layers of edge-sharing octahedra; the layers are stacked in such a way that the halogen atoms, taken by themselves, form a cubic closest-packing in the $CdCl_2$ type and a hexagonal closest-packing in the CdI_2 type. The metal atoms occupy octahedral interstices of the sphere packing. Attention is focused in chapter 15 on the linking of the polyhedra and on the corresponding chemical compositions, while the packing of the molecules or ions in the crystal is a secondary aspect. In this chapter we develop the same facts from the point of view of the general packing. We restrict the discussion mainly to the most important packing principle, that of the closest packings of spheres.

That which applies to the cadmium halides also applies to many other compounds: a proportion of the atoms, taken by themselves, form a closest packing, and the remaining atoms occupy interstices in this packing. The atoms forming the sphere packing do not have to be the same, but they must have similar sizes, in the sense outlined in the sections 14.1 and 14.2 concerning compounds with closest-packed atoms. In perovskite, $CaTiO_3$, for example, the calcium and the oxygen atoms together form a cubic closest-packing, and the titanium atoms occupy certain octahedral interstices. Due to the space requirements of the atoms in the interstices and due to their bonding interactions with the surrounding atoms, the sphere packing frequently experiences certain distortions, but these are often surprisingly small. Moreover, it is possible to include atoms that are too large for the interstices if the packing is expanded; strictly speaking, the packing is then no longer a closest packing (the spheres have no contact with each other), but their relative arrangement in principle remains unchanged.

16.1   The Interstices in Closest Packings of Spheres

Octahedral Interstices in the Hexagonal Closest-packing

Fig. 115(a) shows a section of two superimposed hexagonal layers in a closest packing of spheres. This representation has the disadvantage that the spheres of layer A are largely concealed by the layer B. In all the following figures we will therefore use the representation shown in Fig. 115(b); it shows exactly the same section of the sphere packing, but the spheres are drawn smaller. Of course, since the real size of the spheres is larger, the points of contact between the spheres can no longer be perceived, but we now gain an excellent impression of the sites of the octahedral interstices in the sphere packing: they appear as the

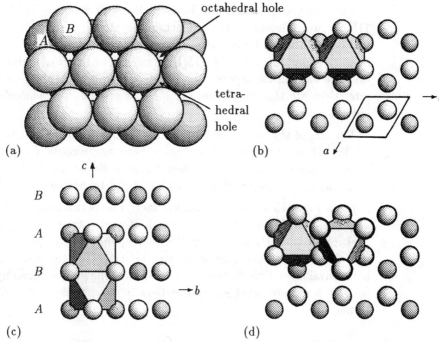

Fig. 115

(a) Relative position of two hexagonal layers in a closest packing of spheres. (b) The same layers with spheres drawn to a smaller scale; two edge-sharing octahedra and the unit cell of the hexagonal closest-packing are shown. (c) Side view of the hexagonal closest-packing; two face-sharing octahedra are shown. (d) Two vertex-sharing octahedra in the hexagonal closest-packing

large holes surrounded by six spheres. The edges of two octahedra are plotted in Fig. 115(b); these two octahedra share a common edge. Fig. 115(c) represents a side view of a hexagonal closest-packing (looking towards the edges of the hexagonal layers); the two plotted octahedra share a common face. The two octahedra shown in Fig. 115(d) are next to each other at different heights, they share a common vertex.

From Fig. 115 we can see how adjacent octahedra are linked in a hexagonal closest-packing:

- face-sharing when the octahedra are located one on top of the other in the direction c;
- edge-sharing when they are adjacent in the a–b plane;
- vertex-sharing when they are adjacent at different heights.

The bond angles at the bridging atoms in the common octahedron vertices are fixed by geometry (angles M–X–M, M in the octahedron centers):

70.5° for face-sharing;
90.0° for edge-sharing;
131.8° for vertex-sharing.

The number of octahedral holes in the unit cell can be deduced from Fig. 115(c): two differently oriented octahedra alternate in direction c, i.e. it takes two octahedra until the pattern is repeated. Hence, there are two octahedral interstices per unit cell. Fig. 115(b) shows the presence of two spheres in the unit cell, one each in the layers A and B. The number of spheres and of octahedral interstices are thus the same, i.e. *there is exactly one octahedral interstice per sphere.*

The size of the octahedral interstices follows from the construction of Fig. 15 (p. 38). There, it is assumed that the spheres are in contact with one another just as in a sphere packing. In the hole between six octahedrally arranged spheres with radius 1 a sphere with radius 0.414 can be accommodated.

From Fig. 115 we realize another fact. The octahedron centers are arranged in planes parallel to the a–b plane, half-way between the sphere layers. The position of the octahedron centers corresponds to the position C which does not occur in the stacking sequence $ABAB$... of the spheres. We designate octahedral interstices in this position in the following sections by γ. By analogy, we will designate octahedral interstices in the positions A and B by α and β, respectively.

Tetrahedral Interstices in the Hexagonal Closest-packing

Fig. 116 shows sections of the hexagonal closest-packing in the same manner as in Fig. 115, but displaying tetrahedra made up of four spheres each. The tetrahedra share vertices in the a–b plane. In the stacking direction pairs of tetrahedra share a common face, and the pairs are connected with each other by common vertices. A pair can be regarded as a trigonal bipyramid. The center of the trigonal bipyramid is identical with the interstice between three atoms in the hexagonal layer; from the center, the axial atoms of the bipyramid are 41 % more distant than the equatorial atoms. If we only consider the three equatorial atoms, the interstice is triangular; if we also take into account the axial atoms, it is trigonal-bipyramidal. The tetrahedral interstices are situated above and below of this interstice. Within a pair of layers AB a tetrahedron pointing upwards shares edges with three tetrahedra pointing downwards.

Fig. 116
Tetrahedra in hexagonal closest-packing: (a) view of the hexagonal layers; (b) view parallel to the hexagonal layers (stacking direction upwards)

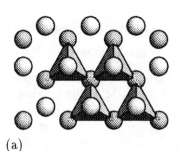

(a)

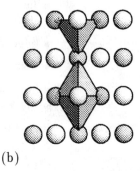

(b)

The bond angles M–X–M at the bridging atoms between two occupied tetrahedra are:

56.7° for face-sharing;
70.5° for edge-sharing;
109.5° for vertex-sharing.

As can be seen from Fig. 116(b), there is one tetrahedral interstice above and one below every sphere, i.e. *there are two tetrahedral interstices per sphere.*

According to the calculation of Fig. 15 (p. 38), a sphere with radius 0.225 fits into the tetrahedral hole enclosed by four spheres of radius 1.

Octahedral and Tetrahedral Interstices in the Cubic Closest-packing

In cubic closest-packing, consideration of the face-centered unit cell is a convenient way to get an impression of the arrangement of the interstices. The octahedral interstices are situated in the center of the unit cell and in the middle of each of its edges (Fig. 117(a)). The octahedra share vertices in the three directions parallel to the unit cell edges. They share edges in the directions diagonal to the unit cell faces. There are no face-sharing octahedra.

If we consider the unit cell to be subdivided into eight octants, we can perceive one tetrahedral interstice in the center of every octant (Fig. 117(b)). Two tetrahedra share an edge when their octants have a common face. They share a vertex if their octants only have a common edge or a common vertex. There are no face-sharing tetrahedra.

There are four spheres, four octahedral interstices and eight tetrahedral interstices per unit cell. Therefore, their numerical relations are the same as for hexagonal closest-packing, as well as for any other stacking variant of closest packings: one octahedral and two tetrahedral interstices per sphere. Moreover, the sizes of these interstices are the same in all closest packings of spheres.

Fig. 117
Face-centered
unit cell of cubic
closest-packing,
(a) with
octahedral
interstices,
(b) with
tetrahedral
interstices

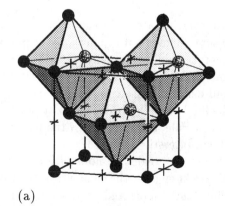

(a)

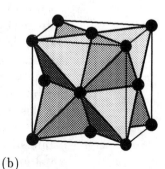

(b)

The bond angles M–X–M at the bridging atoms between two polyhedra occupied by M atoms are:

edge-sharing octahedra	90.0°
vertex-sharing octahedra	180.0°
edge-sharing tetrahedra	70.5°
vertex-sharing tetrahedra in octants with a common edge	109.5°
vertex-sharing tetrahedra in octants with a common vertex	180.0°

The hexagonal layers with the stacking sequence $ABCABC$... are perpendicular to the space diagonals of the unit cell. The layers in one pair of layers, say AB, have the same mutual arrangement as in Fig. 115(b). The position of the following layer C is situated exactly above of the octahedral interstices between A and B. The pattern of the edge-sharing octahedra within *one pair of layers* is independent of the stacking sequence. The sequence of the positions of the octahedron centers in the stacking direction is $\gamma\gamma$... in the hexagonal closest-packing and it is $\gamma\alpha\beta\gamma\alpha$... in the cubic closest-packing (Fig. 118).

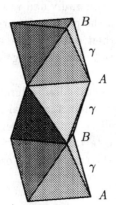 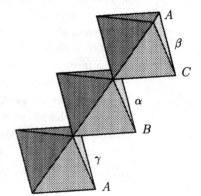

Fig. 118
Relative arrangement of the octahedra in hexagonal and in cubic closest-packing in the direction of stacking of the hexagonal layers

16.2   Interstitial Compounds

The conception of intercalating atoms in the interstices of a sphere packing is not just an idea; with some elements it can actually be performed. The uptake of hydrogen by certain metals yielding metal hydrides is the most familiar example. During the absorption of the hydrogen the properties of the metal experience significant changes and usually phase transitions take place, i.e. the packing of the metal atoms in the final metal hydride is usually not the same as that of the pure metal. However, as a rule, it still is one of the packings typical for metals. For this reason the term interstitial hydrides has been coined. The hydrogen content is variable, we have to deal with nonstoichiometric compounds.

Interstitial hydrides are known for the elements of the third to fifth transition metal groups (including lanthanoids and actinoids), and for chromium, nickel and palladium. Magnesium hydride can also be included, since it can take up hydrogen under pressure up to the composition MgH_2; upon heating the hydrogen is released. The limiting composition is MH_3 for most of the lanthanoids and actinoids, otherwise it is MH_2, except for Ta, Ni, and Pd ($TaH_{0.9}$, $NiH_{0.6}$, and $PdH_{0.83}$). In some cases the compounds are unstable for certain composition ranges (e.g. only cubic $HoH_{1.95}$ to $HoH_{2.24}$ and hexagonal $HoH_{2.64}$ to $HoH_{3.00}$ are stable).

The typical structure for the composition MH_2 is a cubic closest-packing of metal atoms in which all tetrahedral interstices are occupied by H atoms; this is the CaF_2 type. The surplus hydrogen in the lanthanoid hydrides MH_2 to MH_3 is placed in the octahedral interstices (Li_3Bi type for LaH_3 to NdH_3, cf. Fig. 84, p. 150).

The interstitial hydrides of transition metals differ from the salt-like hydrides of the alkali and akaline-earth metals MH and MH_2, respectively, as can be seen from their densities. While the latter have higher densities than the metals, the transition metal hydrides have expanded metal lattices. Furthermore, the transition metal hydrides exhibit metallic luster and are semiconducting. Alkali hydrides have NaCl structure; MgH_2 has rutile structure.

The packing density of the H atoms is very high in all hydrogen-rich metal hydrides. For instance, in MgH_2 it is 55 % higher than in liquid hydrogen. Since magnesium and the alloy $LaNi_5$ take up and release hydrogen relatively easily, they are being considered as potential reservoirs for the storage of hydrogen.

The carbides and nitrides of the elements Ti, Zr, Hf, V, Nb, Ta, Cr, Mo, W, Th, and U are considered to be typical interstitial compounds. Their composition frequently corresponds to one of the approximate formulas M_2X or MX. As a rule, they are nonstoichiometric compounds with compositions ranging within certain limits. This fact, the limitation to a few similar structure types, and very similar properties show that the geometric packing conditions of the atoms have fundamental importance in this class of compounds.

The nitrides can be prepared by heating a metal powder in an N_2 or NH_3 atmosphere to temperatures above 1100 °C. The carbides form upon heating mixtures of the metal powders with carbon to temperatures of about 2200 °C. Both the nitrides and carbides can also be made by chemical transport reactions by the VAN ARKEL–DE BOER method if the metal deposition takes place in an atmosphere of N_2 or a hydrocarbon. Their remarkable properties are:

- Very high hardness with values of 8 to 10 on the MOHS scale; in some cases they approach the hardness of diamond (e.g. W_2C).

- Extremely high melting points, for example (values in °C):

(Ti	1660)	TiC	3140	TiN	2950		
(Zr	1850)	ZrC	3530	ZrN	2980		
(Hf	2230)	HfC	3890	HfN	3300	TaC	3880

(Further values for comparison: melting point of W 3420 °C (highest melting metal), sublimation point of graphite approx. 3350 °C).

- Metallic electrical conductivity, in some cases also superconductivity at low temperatures (e.g. NbC, transition temperature 10.1 K).

- High chemical resistance, except to oxidizing agents at high temperatures (such as atmospheric oxygen above of 1000 °C or concentrated HNO_3).

The intercalation of C or N into the metal thus involves an increased refractoriness with preservation of metallic properties.

The structures can be considered as metal atom packings which have incorporated the nonmetal atoms in their interstices. Usually, the metal atom packings are not the same as those of the corresponding pure metals. The following structure types have been observed:

M_2C and M_2N	hexagonal closest-packing of M atoms, C or N atoms in half of the octahedral interstices
MC and MN	cubic closest-packing of M atoms, C or N atoms in all octahedral interstices = NaCl type (not for Mo, W)
MoC, MoN, WC, WN	WC type

In the WC type the metal atoms do not have a closest packing, but a hexagonal-primitive sphere packing; the metal atoms form trigonal prisms that are occupied by the C atoms.

For the structures of M_2C and M_2N the question arises: is there an ordered distribution of occupied and unoccupied octahedral holes? There are several possibilities for an ordered distribution, some of which actually occur. For example, in W_2C occupied and unoccupied octahedral holes alternate in layers; this is the CdI_2 type. In β-V_2N there are alternating layers in which the octahedral holes are one-third and two-thirds occupied. The question of ordered distributions of occupied interstices is the subject of the following sections.

16.3   Important Structure Types with Occupied Octahedral Interstices in Closest Packings of Spheres

We focus attention here on the binary compounds MX_n, the X atoms being arranged in a closest-packed manner and the M atoms occupying the octahedral interstices. Since the number of octahedral interstices coincides with the number of X atoms, exactly the fraction $1/n$ of them has to be occupied to ensure the correct stoichiometry. As outlined above, in the following we denote the positions of the layers of X atoms by A, B and C, and the intermediate planes of octahedral interstices by α (between B and C), β (between C and A) and γ (between A and B). Fractional numbers indicate the fraction of octahedral holes that are occupied in the corresponding intermediate plane; a completely unoccupied intermediate plane is marked by the SCHOTTKY symbol □.

Compounds MX

structure type	stacking sequence	examples
NaCl	$A\gamma B\alpha C\beta$	LiH, KF, AgCl, MgO, PbS, TiC, CrN
NiAs	$A\gamma B\gamma$	TiS, CoS, CoSb, AuSn

In both the NaCl and the NiAs structure types all octahedral interstices are occupied in a cubic closest-packing or hexagonal closest-packing, respectively. The coordination number is 6 for all atoms. In the NaCl type all atoms have octahedral coordination, and it does not matter whether the structure is regarded as a sphere packing of Na^+ ions with intercalated Cl^- ions or vice versa. The situation is different for the NiAs type; only the arrangement of the As atoms is that of a closest packing, while the nickel atoms in the octahedral interstices (γ positions) are stacked one on top the other (Fig. 119). Only the nickel atoms have octahedral coordination; the coordination polyhedron of the arsenic atoms is a trigonal prism. The structure can also be considered as a primitive hexagonal lattice of Ni atoms; in this lattice the only occurring polyhedra are trigonal prisms, their number being twice the number of the Ni atoms. One half of these prisms are occupied by As atoms (cf. also Fig. 102, p. 172).

The above-mentioned examples show that the NaCl type occurs preferentially in salt-like (ionic) compounds, some oxides and sulfides, and the interstitial compounds discussed in the preceding section. Due to electrostatic reasons the NaCl type is well-suited for very polar compounds, since every atom has only atoms of the other element as closest neighbors. Sulfides, selenides, and tellurides, as well as phosphides, arsenides and antimonides, with NaCl structure have been observed with alkaline earth metals and with elements of the third transition metal group (MgS, CaS, ..., MgSe, ..., BaTe; ScS, YS, LnS, LnSe, LnTe; LnP, LnAs, LnSb with Ln = lanthanoid). However, with other transition metals, the NiAs type and its distorted variants are preferred; this is less favorable than the NaCl type electrostatically since the Ni atoms in the face-sharing octahedra are rather close to each other (Ni–Ni distance 252 pm, just a little longer than the Ni–As distance of 243 pm). This suggests the presence of bonding metal–metal interactions, particularly since this structure type only occurs if the metal atoms still have d electrons available. The existence of metal–metal interactions is also supported by the following observations: metallic luster and conductivity, variable composition, and the dependence of the lattice parameters on the electronic configuration, e.g.:

ratio c/a of the hexagonal unit cell

TiSe	VSe	CrSe	$Fe_{1-x}Se$	CoSe	NiSe
1.68	1.67	1.64	1.64	1.46	1.46

Even smaller ratios c/a are observed for the more electron-rich arsenides and antimonides (e.g. 1.39 for NiAs). Since the ideal c/a ratio of hexagonal closest-packing is 1.633, there is a considerable compression in the c direction, i.e. in the direction of the closest contacts among the metal atoms.

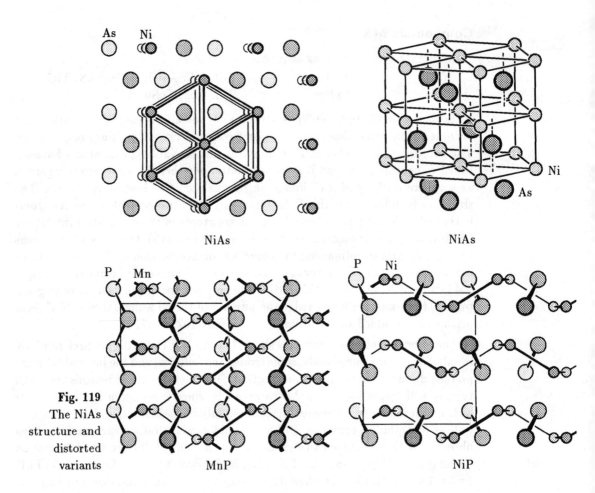

Fig. 119
The NiAs
structure and
distorted
variants

The structure of MnP is a distorted variety of the NiAs type: every metal atom also has close contacts in a zigzag line parallel to the a–b plane, which amounts to a total of four close metal atoms (Fig. 119). Simultaneously, the P atoms have moved up to a zigzag line; this can be interpreted as a $(P^-)_\infty$ chain in the same manner as in ZINTL phases. In NiP the distortion is different, allowing for the presence of P_2 pairs (P_2^{4-}). These distortions are to be taken as PEIERLS distortions. Calculations of the electronic band structures can be summarized in short: 9–10 valence electrons per metal atom favor the NiAs structure, 11–14 the MnP structure, and more than 14 the NiP structure (phosphorus contributes 5 valence electrons per metal atom); this is valid for phosphides. Arsenides and especially antimonides prefer the NiAs structure also for the larger electron counts.

Compounds which have the NiAs structure often exhibit a certain phase width in that metal atom positions can be vacant. The composition then is $M_{1-x}X$. The vacancies can have a random or an ordered distribution. In the latter case we have to deal with superstructures of the NiAs type; they are known, for example, among iron sulfides such as Fe_9S_{10} or $Fe_{10}S_{11}$. If metal atoms are removed from every other layer, we have a continuous series from $M_{1.0}X$ with the NiAs structure down to $M_{0.5}X$ ($= MX_2$) with the CdI_2 structure; phases of this kind are known for $Co_{1-x}Te$ (CoTe: NiAs type; $CoTe_2$: CdI_2 type).

Compounds MX_2

In compounds MX_2, one half of the octahedral interstices are occupied. There are several possibilities for the distribution of occupied and vacant interstices in the intermediate planes:

1. Fully occupied and vacant intermediate planes alternate. In the occupied planes the octahedra share common edges (Fig. 95, p. 166).

structure type	stacking sequence	examples
$CdCl_2$	$A\gamma B\square C\beta A\square B\alpha C\square$	$MgCl_2$, $FeCl_2$, Cs_2O
CdI_2	$A\gamma B\square$	$MgBr_2$, PbI_2, SnS_2, $Mg(OH)_2$, $Cd(OH)_2$, Ag_2F

In addition, further polytypes exist, i.e. structures having other stacking sequences of the halogen atoms. Especially for CdI_2 itself a large number of such polytypes are now known; for this reason the term CdI_2 type is nowadays considered unfortunate, and the terms $Mg(OH)_2$ (brucite) or $Cd(OH)_2$ type are preferred by some authors. The H atoms of the hydroxides are oriented into the tetrahedral interstices in between the layers, and do not act as H bridges. Botallackite, $Cu_2(OH)_3Cl$, has a structure like CdI_2, every other layer of the sphere packing consisting of Cl atoms and OH groups (another modification with this composition is atacamite, which is mentioned below).

2. The intermediate planes are alternately two-thirds and one-third occupied.

structure type	stacking sequence	examples
ε-Fe_2N	$A\gamma_{2/3}B\gamma'_{1/3}$	β-Nb_2N, Li_2ZrF_6

The intermediate planes with two-thirds occupation have octahedra sharing edges with the honeycomb pattern as in BiI_3; the octahedra in the intermediate planes with one-third occupation are not directly connected with one another, but they have common vertices with octahedra of the adjacent layers. In Li_2ZrF_6 the Zr atoms are those in the intermediate plane with one-third occupation (cf. Fig. 99, p. 169).

3. The intermediate planes are alternately one-quarter and three-quarters occupied. This is the arrangement in atacamite, a modification of $Cu_2(OH)_3Cl$ with the stacking sequence:

$$A\gamma_{1/4}B\alpha_{3/4}C\beta_{1/4}A\gamma'_{3/4}B\alpha'_{1/4}C\beta'_{3/4}.$$

4. Every intermediate plane is half occupied.

structure type	stacking sequence	examples
$CaCl_2$	$A\gamma_{1/2}B\gamma'_{1/2}$	$CaBr_2$, ε-$FeO(OH)$, Co_2C
α-PbO_2	$A\gamma_{1/2}B\gamma''_{1/2}$	TiO_2 (high pressure)
α-$AlO(OH)$	$A\gamma_{1/2}B\gamma'''_{1/2}$	α-$FeO(OH)$ (goethite)

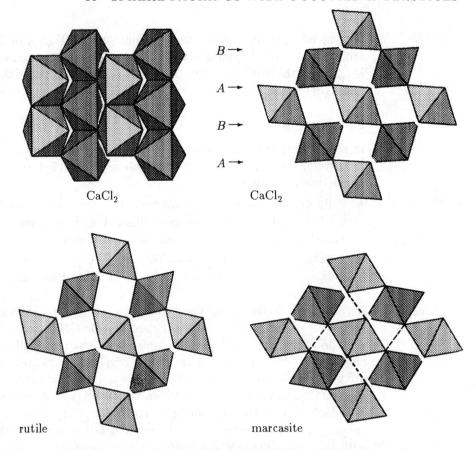

Fig. 120
Top left: CaCl$_2$
structure, view
perpendicular to
the hexagonal
layers. Top right
and bottom: view
along the strands
of edge-sharing
octahedra in
CaCl$_2$, rutile and
marcasite; the
hexagonal layers
in CaCl$_2$ are
marked by A and
B. Dashed: S–S
bonds

CaCl$_2$ CaCl$_2$

rutile   marcasite

In CaCl$_2$ linear strands of edge-sharing octahedra are present, and the strands are joined by common octahedron vertices (Fig. 120). Marcasite is a modification of FeS$_2$ related to the CaCl$_2$ type, but distorted by the presence of S$_2$ dumbbells. The joining of adjacent S atoms to form dumbbells is facilitated by a mutual rotation of the octahedron strands (Fig. 120). Several compounds adopt this structure type, e.g. NiAs$_2$ and CoTe$_2$.

If the octahedron strands are rotated in the opposite direction, the rutile type results (Fig. 120). Due to this relation with the CaCl$_2$ type, a hexagonal closest-packing of O atoms has frequently been ascribed to the rutile type. However, the deviations from this kind of packing are quite significant. For one thing, every O atom is no longer in contact with twelve other O atoms, but only with eleven; moreover, the "hexagonal" layers are considerably corrugated. By the formalisms of group theory it is also not permissible to regard the tetragonal rutile as a derivative of the hexagonal closest-packing (cf. chapter 19.3). In fact, the arrangement of the O atoms in rutile is that of a tetragonal close-packing of spheres. This is a sphere packing filling space to 71.9 %, which is only slightly less than in a closest packing. Consider a ladder which has spheres at the joints of the rungs (Fig. 121). Set up such ladders vertical, but mutually rotated around the vertical axis in such a way that the spheres of one ladder come to be next

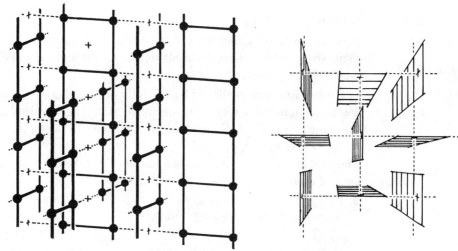

Fig. 121
Tetragonal
close-packing of
spheres

to the gaps between the rungs of the adjacent ladders. In the packing obtained
this way every sphere has coordination number 11: 2+4+2 spheres from three
neighboring ladders and 3 spheres within the ladder. The gaps between the
rungs correspond to the octahedral interstices that are occupied by Ti atoms
in rutile, and the ladders correspond to the strands of edge-sharing octahedra.
Compared to a closest packing of spheres (c.n. 12), the coordination number
is reduced by 8 %, but the space filling is reduced only by 3 %; this gives an
idea of why the rutile type is the preferred packing for highly polar compounds
(dioxides, difluorides).

α-PbO_2 is another structure which has mutually connected strands of edge-
sharing octahdera; the strands have zigzag shape (Fig. 98, p. 168). Linear
strands of edge-sharing octahedra as in $CaCl_2$, but which form edge-sharing
double-strands, are present in diaspore, α-$AlO(OH)$ (Fig. 122).

Fig. 122
Double-strands of edge-sharing octahedra in
diaspore, α-$AlO(OH)$

Compounds MX_3

In compounds MX_3, one third of the octahedral interstices are occupied. Again, there are several possible distributions of vacant and occupied interstices in the intermediate planes:

1. Every third intermediate plane is fully occupied, and the others are unoccupied. The octahedra within the occupied planes share edges as in CdI_2. This is the structure of Cr_2AlC, in which the sequence of layers is:

$$A_{Cr} \gamma B_{Cr} \square A_{Al} \square B_{Cr} \gamma A_{Cr} \square B_{Cr} \square$$

The carbon atoms take positions inside of octahedra made up from only one kind of atom, those of the transition metal.

2. Every other intermediate plane is two-thirds occupied.

structure type	stacking sequence	examples
$AlCl_3$	$A\gamma_{2/3}B \square C\beta_{2/3}A \square B\alpha_{2/3}C \square$	YCl_3, $CrCl_3$ (high temp.)
BiI_3	$A\gamma_{2/3}B \square$	$FeCl_3$, $CrCl_3$ (low temp.)

Both structure types have the same kind of layer of edge-sharing octahedra (Fig. 94; cf. also the cover). Among the trihalides with this kind of layer structure, stacking disorder is quite common, i.e. the stacking sequence of the hexagonal halogen atom layers is not strictly AB or ABC, but stacking faults occur frequently. This also applies to $AlCl_3$ and BiI_3 themselves; the frequency of the stacking faults depends on the growth conditions of the specific single crystal. For example, one crystal of BiI_3 that had been obtained by sublimation essentially had the hexagonal stacking sequence $hhh\dots$, but in a random sequence one out of 16 layers was a c layer.

3. Every intermediate layer is one-third occupied.

structure type	stacking sequence	examples
$RuBr_3$	$A\gamma_{1/3}B\gamma_{1/3}$	β-$TiCl_3$, ZrI_3, $MoBr_3$
RhF_3	$A\gamma_{1/3}B\gamma'_{1/3}$	IrF_3, PdF_3

In the $RuBr_3$ type a succession of face-sharing octahedra forms a strand in the c direction. The metal atoms in adjacent octahedra are shifted pairwise from the octahedron centers, forming metal–metal bonds (Fig. 96, p. 167). This seems to be the condition for the existence of this structure type, i.e. it only occurs with transition metals that still have d electrons available.

In the RhF_3 type all octahedra share vertices, and corresponding to the hexagonal closest-packing of the F atoms the Rh–F–Rh angles are approximately 132°. By mutual rotation of the octahedra the angle can be widened up to 180°, but then the packing is less dense. This has been observed for the VF_3 type (V–F–V angle approx. 150°) which occurs with some trifluorides (GaF_3, TiF_3, FeF_3 etc.); cf. Fig. 91, p. 162. In PdF_3 the Pd–F lengths in the octahedra alternate (217 and 190 pm) in accordance with the formula $Pd^{II}Pd^{IV}F_6$.

Compounds M_2X_3

Two thirds of the octahedral interstices are occupied. In a way the possible structure types are the "inverse" of the MX_3 structures, since in these two thirds of the octahedral interstices are vacant. If we take an MX_3 type, clear the occupied interstices and occupy the vacant ones, the result is an M_2X_3 structure. The kind of linking between the occupied octahedra, however, is different. The arrangement of the vacant octahedral interstices of the RhF_3 type corresponds to the occupied interstices in corundum, Al_2O_3; its occupied octahedra share edges and faces (Fig. 101, p. 171). The layer sequence is:

$$A\gamma_{2/3}B\gamma'_{2/3}A\gamma''_{2/3}B\gamma_{2/3}A\gamma'_{2/3}B\gamma''_{2/3}$$

Compounds MX_4, MX_5, and MX_6

$\frac{1}{4}$, $\frac{1}{5}$ and $\frac{1}{6}$ of the octahedral interstices are occupied, respectively. There are various possibilities for the distribution of the occupied sites, and the specification of a layer sequence alone is not very informative. Fig. 123 shows some examples which also allow us to recognize an important principle concerning the packing of molecules: all octahedral interstices that immediately surround a molecule must be vacant, and then occupied interstices have to follow; otherwise either the molecules would be joined to polymer units or not all of the atoms of the sphere packing would be part of a molecule. These statements may seem self-evident; however, they imply a severe restriction on the number of possible packing varieties for a given kind of molecule. For tetrahalides consisting of chains of edge-sharing octahedra, we noted on p. 165 that numerous chain configurations are conceivable. However, because of these packing necessities, some of them are not compatible with a closest packing; no examples are known for them and it is unlikely that any will ever be observed [131].

16.4   Perovskites

In the structure of γ'-Fe_4N the Fe atoms form a cubic closest-packing, and one fourth of the octahedral interstices are occupied with N atoms. A special aspect of the structure concerns the Fe atoms, as there are two different kinds of them. Only one kind, making up three out of four Fe atoms, is a constituent part of the occupied octahedra. The occupied octahedra share vertices and form a three-dimensional network corresponding to the ReO_3 structure (Fig. 91, p. 162). The fourth Fe atom is in the center of the cubic unit cell.

If we substitute the two kinds of Fe atoms of the sphere packing for atoms of two different elements, say oxygen and calcium, and if we occupy those octahedra which are formed solely of the oxygen atoms by titanium atoms, the result is the structure of perovskite, $CaTiO_3$ (Fig. 124). The Ca and O atoms jointly form the cubic closest-packing in an arrangement corresponding to the ordered alloy $AuCu_3$ (Fig. 82, p. 147). The atomic order in the hexagonal layers of the sphere packing is that shown for $AuCu_3$ on p. 148. Being a part of the sphere packing, a Ca^{2+} ion has coordination number 12.

If the position of the Ca^{2+} ion is vacant, the remaining framework is that of the ReO_3 type. The analogy between ReO_3 and $CaTiO_3$ is not a mere formalism, since the partial occupation of the Ca positions with varying amounts of metal

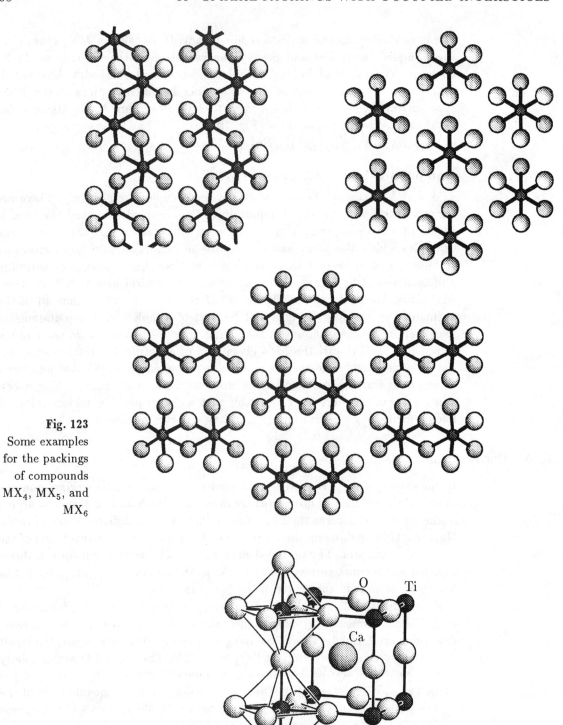

Fig. 123
Some examples
for the packings
of compounds
MX$_4$, MX$_5$, and
MX$_6$

Fig. 124
The perovskite structure

ions can actually be achieved, specifically in the case of the cubic tungsten bronzes, $A_x WO_3$ (A = alkali metal, x = 0.3 to 0.93). Their color and the oxidation state of the tungsten depend on the value of x; they have metallic luster, and with $x \approx 1$ they are gold-colored, with $x \approx 0.6$ red and with $x \approx 0.3$ dark violet.

In normal, cubic perovskite the close-packed hexagonal CaO_3 layers have the stacking sequence $ABC\ldots$ or $c\ldots$ and the occupied octahedra only share vertices. The structural family of perovskites also includes numerous other stacking varieties, with c and h layers in different sequences. At an h layer the octahedra share faces. In a sequence like $chhc$, there is a group of three octahedra that share faces at the h layers; this group is connected with other octahedra by vertex-sharing at the c layers. The size of the groups of face-sharing octahedra depends on the nature of the metal atoms in the octahedra and especially on the ionic radius ratios. Fig. 125 shows some examples.

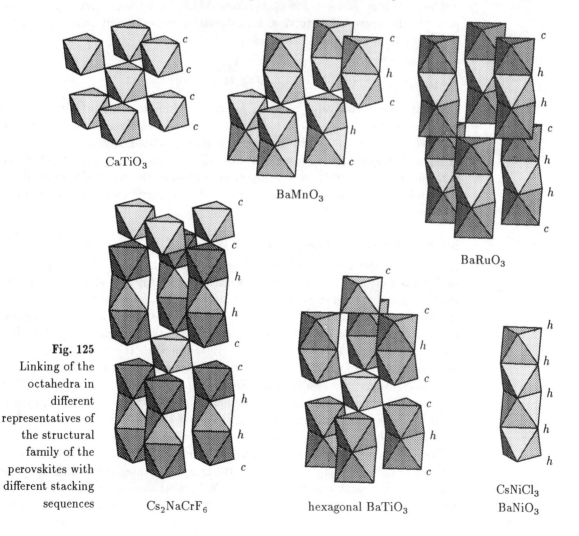

Fig. 125
Linking of the octahedra in different representatives of the structural family of the perovskites with different stacking sequences

CaTiO$_3$

BaMnO$_3$

BaRuO$_3$

Cs$_2$NaCrF$_6$

hexagonal BaTiO$_3$

CsNiCl$_3$
BaNiO$_3$

The ideal, cubic perovskite structure is not very common; even the mineral perovskite itself, $CaTiO_3$, is slightly distorted. $SrTiO_3$ is undistorted. As shown in Fig. 91 (p. 162), the ReO_3 type can be converted to a more dense packing by mutual rotation of the octahedra until a hexagonal closest-packing is obtained in the RhF_3 type. During the rotation the void in the center of the ReO_3 unit cell becomes smaller and finally becomes an octahedral interstice in the sphere packing of the RhF_3 type. If this octahedral interstice is occupied, we have the ilmenite type ($FeTiO_3$). By an appropriate amount of rotation of the octahedra, the size of the hole can be adapted to the size of the A ion in a perovskite. In addition, some tilting of the octahedra allows a variation of the coordination number and coordination polyhedra. Distorted perovskites have reduced symmetry, which is important for the magnetic and electric properties of these compounds. Due to these properties perovskites have great industrial importance, especially the ferroelectric $BaTiO_3$. This is discussed in chapter 17.

The *tolerance factor* t for perovskites AMX_3 is a value that allows us to estimate the degree of distortion. Its calculation is performed using ionic radii, i.e. purely ionic bonding is assumed:

$$t = \frac{r(A) + r(X)}{\sqrt{2}[r(M) + r(X)]}$$

Geometry requires a value of $t = 1$ for the ideal cubic structure. In fact, this structure occurs if $0.89 < t < 1$. Distorted perovskites occur if $0.8 < t < 0.89$. With values less than 0.8, the ilmenite type is more stable (Fig. 101, p. 171). The hexagonal stacking variants such as those in Fig. 125 usually have $t > 1$. Since perovskites are not truly ionic compounds and since the result also depends on which values are taken for the ionic radii, the tolerance factor is only a rough estimate.

Superstructures of the perovskite type

If we enlarge the unit cell of perovskite by doubling all three edges, it is possibile to occupy equivalent positions with atoms of different elements. Fig. 126 shows some representatives of the elpasolite family. In elpasolite, K_2NaAlF_6, the potassium and the fluoride ions jointly form the cubic closest-packing, i.e. K^+ and F^- take the Ca and O positions of perovskite, respectively. The one-to-one relation can be recognized by comparison with the doubled formula of perovskite, $Ca_2Ti_2O_6$. The comparison also shows the partition of the octahedral Ti positions into two sites for Na and Al. In kryolite, Na_3AlF_6, the Na^+ ions occupy two different positions, namely those of Na^+ and K^+ in elpasolite, i.e. positions with coordination numbers of 6 and 12. Since this is not convenient for ions of the same size, the lattice experiences some distortion.

Perovskites which have copper atoms in the octahedral sites and are deficient in oxygen, $ACuO_{3-\delta}$, have been the subject of intense research in recent years. Alkaline earth and trivalent ions (Y^{3+}, lanthanoids, Bi^{3+}, Tl^{3+}) occupy the A positions. A typical composition is $YBa_2Cu_3O_{7-x}$ with $x \approx 0.04$. These

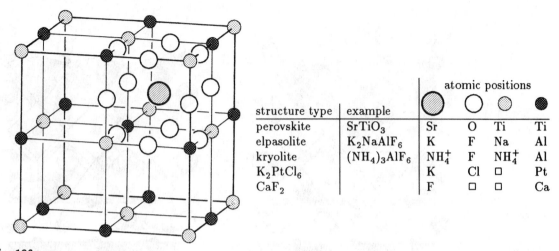

structure type	example	atomic positions			
		⬭	○	◌	●
perovskite	$SrTiO_3$	Sr	O	Ti	Ti
elpasolite	K_2NaAlF_6	K	F	Na	Al
kryolite	$(NH_4)_3AlF_6$	NH_4^+	F	NH_4^+	Al
K_2PtCl_6		K	Cl	□	Pt
CaF_2		F	□	□	Ca

Fig. 126
Superstructures of the perovskite type. Only in one octant have all atoms been plotted; the atoms on the edges and in the centers of all octants are the same

compounds are "high-temperature" superconductors, i.e. they become super-conducting at temperatures as high as the boiling point of N_2 (77 K) or above. The structure is a superstructure of perovskite, but with approximately $\frac{7}{9}$ of the oxygen positions vacant, in such a way that $\frac{2}{3}$ of the Cu atoms have square-pyramidal coordination and $\frac{1}{3}$ have square-planar coordination (Fig. 127). The structures of some other representatives of this compound class are considerably more complicated, and may exhibit disorder and other particularities.

16.5   Occupation of Tetrahedral Interstices in Closest Packings of Spheres

Face-sharing coordination tetrahedra would be present if all tetrahedral interstices in a hexagonal closest-packing were occupied. This would be an unfavorable arrangement for electrostatic reasons. On the other hand, the occupation of all tetrahedral interstices in cubic closest-packing results in an electrostatically favorable structure type: the CaF_2 type (F^- ions in the interstices), which also is the structure of Li_2O (Li^+ in the interstices). The tetrahedra are linked by common edges and common vertices.

Taking out one half of the atoms from the tetrahedral interstices of the CaF_2 type leaves us with the composition MX. Several structure types can arise, depending on the selection of the interstices to be vacated: the sphalerite type with a network of vertex-sharing tetrahedra, the PbO type with layers of edge-sharing tetrahedra, and the PtS type (Fig. 128). In PbO and PtS the metal atoms form the sphere packing. PbO only has tetrahedra occupied by O atoms at the height $z = \frac{1}{4}$, and those in $z = \frac{3}{4}$ are vacant; the Pb atoms at $z \approx 0$ and $z \approx \frac{1}{2}$ and the O atoms together form a layer in which every Pb atom has square-pyramidal coordination (cf. Fig. 114, p. 183). The distribution of the atoms in PtS results in a planar coordination around a Pt atom. The packing is

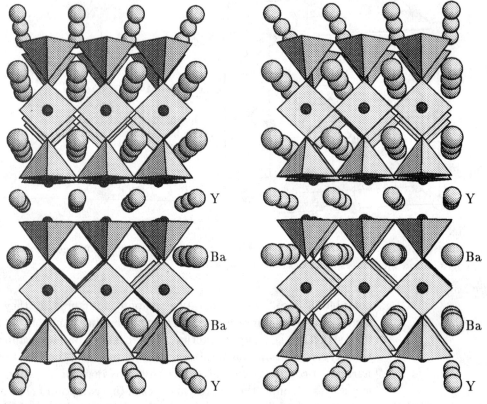

Fig. 127 Structure of $YBa_2Cu_3O_7$. The perovskite structure is attained by inserting O atoms in between the strings of Y atoms and in between the CuO_4 squares (stereo image)

a compromise between the requirements of tetrahedral coordination for sulfur and square coordination for platinum. With a ratio $c/a = 1.00$ there would be an ideal sphere packing with tetrahedral bond angles at S, but rectangular coordination at Pt; with $c/a = 1.41$ the bond angles would be 90° at Pt, but also at S. The actual ratio is $c/a = 1.24$.

HgI_2 and α-$ZnCl_2$ are examples of structures with cubic closest-packing of halogen atoms, having one quarter of the tetrahedral interstices occupied. These tetrahedra share vertices, every tetrahedron vertex being common to two tetrahedra with bond angles of 109.5° at the bridging atoms. The HgI_2 structure corresponds to a PbO structure in which half of the O atoms have been removed and cations have been exchanged with anions (Fig. 128). There are layers, and all Hg atoms are at the same level within one layer (cf. also Fig. 109, p. 179). If half of the atoms are removed from sphalerite, in the way shown in the right part of Fig. 128, the result is the α-$ZnCl_2$ structure. It has a framework of mutually linked helices. The c axis is doubled and the Zn atoms form helices in the c direction. By mutually rotating the tetrahedra, the lattice is widened and the bond angles at the bridging atoms become larger; the result is the cristobalite structure (Fig. 129; the axes of the helices are marked by the symbols ✦. See also Fig. 59, p. 112).

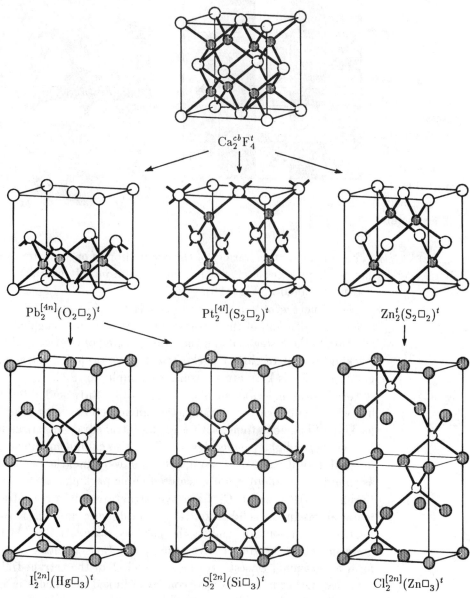

Fig. 128

Relationships among the structures of CaF_2, PbO, PtS, ZnS, HgI_2, SiS_2, and α-$ZnCl_2$. In the top row all tetrahedral interstices (= centers of the octants of the cube) are occupied. Every arrow designates a step in which the number of occupied tetrahedral interstices is halved; this includes a doubling of the unit cells in the bottom row. Light hatching = metal atoms, dark hatching = non-metal atoms. The atoms given first in the formulas form the cubic closest-packing

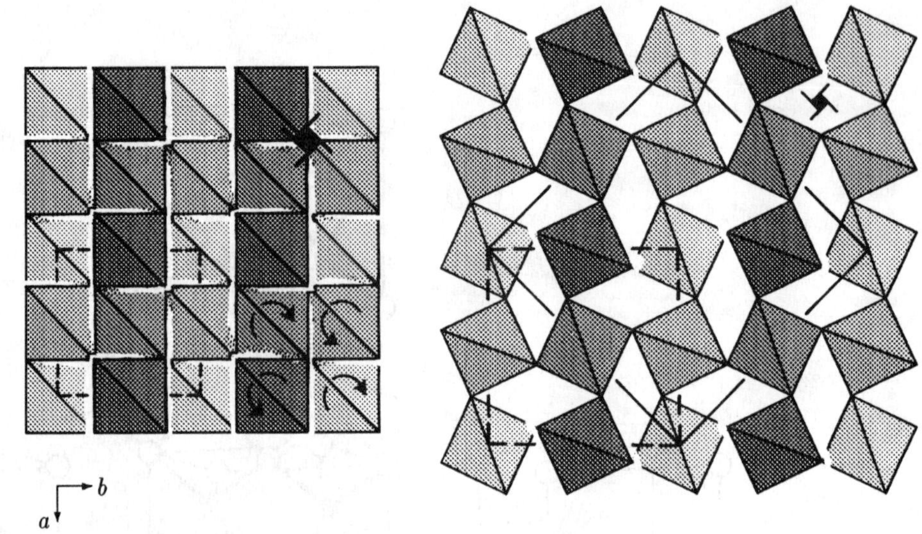

Fig. 129
By mutual
rotation of the
tetrahedra the
α-ZnCl$_2$
structure is
converted to the
cristobalite
structure

SiS_2 offers another variety of the occupation of one quarter of the tetrahedral interstices in a cubic closest-packing of S atoms. It contains strands of edge-sharing tetrahedra (Fig. 128).

The structure of wurtzite corresponds to a hexagonal closest-packing of S atoms in which half of the tetrahedral interstices are occupied by Zn atoms. In addition to the hexagonal and the cubic packing of the two ZnS types, any other stacking variant of closest packings can have occupied tetrahedral interstices; polytypes of this kind are known, for example, for SiC.

Tetrahedral molecules such as $SnCl_4$, $SnBr_4$, SnI_4, and $TiBr_4$ usually crystallize with a cubic closest-packing, though sometimes with a hexagonal closest-packing of halogen atoms, with $\frac{1}{8}$ of the tetrahedral interstices being occupied. Especially the lighter molecules like CCl_4 also exhibit modifications which have molecules rotating in the crystal; averaged over time, the molecules then appear as spheres, and adopt a body-centered cubic packing.

Whereas $AlCl_3$ and $FeCl_3$ have layer structures with octahedrally coordinated metal atoms in the solid state, they form dimeric molecules (two edge-sharing tetrahedra) in solution and in the gaseous state. Al_2Br_6, Al_2I_6 and Ga_2Cl_6, however, retain the dimeric structure even in the solid state. The halogen atoms form a hexagonal closest-packing in which $\frac{1}{6}$ of the tetrahedral interstices are occupied. Other molecules that consist of linked tetrahedra in many cases also are packed in the solid state according to the principle of closest-packings of spheres with occupied tetrahedral interstices, for example Cl_2O_7 or Re_2O_7.

16.6 Spinels

Sphere packings having occupied tetrahedral and octahedral interstices usually occur if atoms of two different elements are present, one of which prefers tetrahedral coordination, and the other octahedral coordination with the atoms of the sphere packing. This is a common feature among silicates (cf. section 15.7). Another important structure type of this kind is the spinel type. Spinel is the mineral $MgAl_2O_4$, and generally spinels have the composition AM_2X_4. Most of them are oxides; in addition, there exist sulfides, selenides, halides und pseudohalides.

In the following, we start by assuming purely ionic structures. In spinel the oxide ions form a cubic closest-packing. Two thirds of the metal ions occupy octahedral interstices, the rest tetrahedral ones. In a "normal" spinel the A ions are found in the tetrahedral interstices and the M ions in the octahedral interstices; we express this by the subscripts T and O, e.g. $Mg_T[Al_2]_OO_4$. Since tetrahedral holes are smaller than octahedral holes, the A ions should be smaller than the M ions. Remarkably, this condition is not fulfilled in many spinels, and just as remarkable is the occurrence of "inverse" spinels which have half of the M ions occupying tetrahedral sites and the other half occupying octahedral sites while the A ions occupy the remaining octahedral sites. Table 22 summarizes these facts and also includes a classification according to the oxidation states of the metal ions.

Arbitrary intermediate states also exist between normal and inverse spinels; they can be characterized by the *degree of inversion* λ:

$\lambda = 0$: normal spinel $\lambda = 0.5$: inverse spinel

The distribution of the cations among the tetrahedral and octahedral sites is then expressed in the following way: $(Mg_{1-2\lambda}Fe_{2\lambda})_T[Mg_{2\lambda}Fe_{2(1-\lambda)}]_OO_4$. The value of λ is temperature-dependent. For example, at room temperature $MgFe_2O_4$ has $\lambda = 0.45$ and thus is essentially inverse.

The difficulties in understanding the cation distributions and in explaining the occurrence of inverse spinels on the basis of ionic radii shows how insufficient this kind of approach is. A somewhat better approach considers the electrostatic part of the lattice energy, the calculated MADELUNG constant being a useful quantity. For a II,III-spinel which has an undistorted sphere packing the MADELUNG constant of the normal spinel is 1.6 % smaller than that of the inverse one, i.e. the inverse distribution is slightly more favorable. However, small distortions that commonly occur in most spinels (widening of the tetrahedral

Table 22 Synopsis of spinel types with examples

oxidation state combination	"normal" spinels $A_T[M_2]_OX_4$	"inverse" spinels $M_T[AM]_OX_4$
II, III	$MgAl_2O_4$	$MgIn_2O_4$
II, III	Co_3O_4	Fe_3O_4
IV, II	$GeNi_2O_4$	$TiMg_2O_4$
II, I	$ZnK_2(CN)_4$	$NiLi_2F_4$
VI, I	WNa_2O_4	

Ionic radii:	Mg^{2+} 72 pm	Fe^{2+} 78 pm	Co^{2+} 75 pm
	Al^{3+} 54 pm	Fe^{3+} 65 pm	Co^{3+} 61 pm

Table 23 Ligand field stabilization energies for Mn_3O_4, Fe_3O_4 and Co_3O_4. Values for high-spin complexes in all cases except for octahedral Co^{III}

	normal	inverse
	$Mn_T^{II}[Mn_2^{III}]_O O_4$	$Mn_T^{III}[Mn^{II}Mn^{III}]_O O_4$
Mn^{II}	0	0
Mn^{III}	$2 \times 0.6 = 1.2$	$0.18 + 0.6 = 0.78$
	$1.2 \Delta_O$	$0.78 \Delta_O$
	$Fe_T^{II}[Fe_2^{III}]_O O_4$	$Fe_T^{III}[Fe^{II}Fe^{III}]_O O_4$
Fe^{II}	0.27	0.40
Fe^{III}	0	0
	$0.27 \Delta_O$	$0.40 \Delta_O$
	$Co_T^{II}[Co_2^{III}]_O O_4$	$Co_T^{III}[Co^{II}Co^{III}]_O O_4$
Co^{II}	0.53	0.80
Co_T^{III} h.s.		0.27
Co_O^{III} l.s.	$2 \times 2.4 = 4.80$	2.40
	$5.33 \Delta_O$	$3.47 \Delta_O$

interstices) can reverse this. In fact, spinels are not purely ionic compounds and the consideration of electrostatic interactions alone is hardly adequate, although it does work quite well for spinels of main group elements. With transition metals, in addition, the considerations of ligand field theory have to be taken into account. To illustrate this, we take as examples the spinels Mn_3O_4, Fe_3O_4 and Co_3O_4. Except for Co(III) these are made up of high-spin complexes. The relative ligand field stabilization energies are, expressed as multiples of Δ_O (cf. Table 11, p. 67):

Mn_O^{2+}	0	Fe_O^{2+}	$\frac{2}{5} = 0.4$	Co_O^{2+}	$\frac{4}{5} = 0.8$	
Mn_T^{2+}	0	Fe_T^{2+}	$\frac{3}{5} \cdot \frac{4}{9} = 0.27$	Co_T^{2+}	$\frac{6}{5} \cdot \frac{4}{9} = 0.53$	
Mn_O^{3+}	$\frac{3}{5} = 0.6$	Fe_O^{3+}	0	Co_O^{3+}	$\frac{12}{5} = 2.4$ (low-spin)	
Mn_T^{3+}	$\frac{2}{5} \cdot \frac{4}{9} = 0.18$	Fe_T^{3+}	0	Co_T^{3+}	$\frac{3}{5} \cdot \frac{4}{9} = 0.27$	

$\Delta_T = \frac{4}{9} \Delta_O$ was taken for tetrahedral ligand fields. Mn_3O_4 is a normal spinel, $Mn_T^{II}[Mn_2^{III}]_O O_4$. If it were to be converted to an inverse spinel, half of the Mn^{III} atoms would have to shift from the octahedral to the tetrahedral environment, and this would imply a decreased ligand field stabilization for these atoms (Table 23); for the Mn^{II} atoms the shifting would make no difference. Fe_3O_4 is an inverse spinel, $Fe_T^{III}[Fe^{II}Fe^{III}]_O O_4$. For the Fe^{III} atoms the exchange of positions would make no difference; for the Fe^{II} atoms it would be unfavorable ($0.4 \Delta_O \rightarrow 0.27 \Delta_O$).

In the case of Co_3O_4, which is a normal spinel, $Co_T^{II}[Co_2^{III}]_O O_4$, the situation is different because octahedrally coordinated Co^{III} almost never occurs in high-spin complexes (its d^6 configuration corresponds to the maximum of ligand field stabilization energy in the low-spin state). If Co_O^{3+} would adopt a high-spin state in Co_3O_4, this should be an inverse spinel. However, in the low-spin

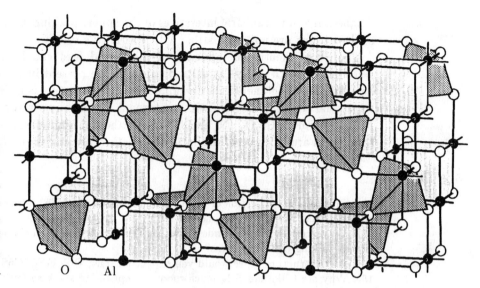

Fig. 130
The spinel
structure (two
unit cells). The
Mg^{2+} ions are
located in the
centers of the
dark tetrahedra

state the normal spinel is favored (Table 23). In addition, the ionic radius does also have an effect; it decreases in the series $Mn^{2+}–Fe^{2+}–Co^{2+}–Ni^{2+}–Cu^{2+}–Zn^{2+}$ and therefore favors the tetrahedral coordination towards the end of the series. Co^{2+} generally shows a tendency towards tetrahedral coordination in its compounds. In Fig. 26 (p. 68) this influence of the ionic size is taken into account by having the dashed line bent for octahedral coordination; this line corresponds to the notional reference state (spherical distribution of the d electrons), relative to which the ligand field stabilization energy is defined. According to Fig. 26, Co^{2+} is more stable in a tetrahedral environment.

Fig. 130 shows the spinel structure. There are four Al^{3+} and four O^{2-} ions in the vertices of an Al_4O_4 cube. Every Al^{3+} ion belongs to two such cubes, so that every cube is linked with four more cubes and every Al^{3+} ion has octahedral coordination. In addition, every O^{2-} ion belongs to an MgO_4 tetrahedron. Each of these tetrahedra shares vertices with four cubes. The cubic unit cell contains eight MgO_4 tetrahedra and eight Al_4O_4 cubes. The metal ions, taken by themselves, have the same arrangement as in the cubic LAVES phase $MgCu_2$ (cf. Fig. 85, p. 153).

The coordination of an O^{2-} ion is three Al^{3+} ions within an Al_4O_4 cube and one Mg^{2+} ion outside of this cube. This way it fulfils the electrostatic valence rule (PAULING's second rule, cf. p. 44), i.e. the sum of the electrostatic bond strengths of the cations corresponds exactly to the charge on an O^{2-} ion:

$$z(O) = -(\ \underbrace{3 \cdot \frac{3}{6}}_{3\,Al^{3+}} + \underbrace{1 \cdot \frac{2}{4}}_{1\,Mg^{2+}}\ ) = -2$$

The required local charge balance between cations and anions which is expressed in PAULING's rule causes the distribution of cations and anions among the octahedral and tetrahedral interstices of the sphere packing. Other distributions of the cations are not compatible with PAULING's rule.

The above-mentioned influence of the ligand field can be discerned when metal atoms with JAHN–TELLER distortions are present in a spinel. Mn_3O_4 is an example: its octahedra are elongated, and the structure is no longer cubic but tetragonal. Other examples with tetragonal distortions are the normal spinels $NiCr_2O_4$ and $CuCr_2O_4$ (Ni and Cu in tetrahedral interstices); in $NiCr_2O_4$ the tetrahedra are elongated, and in $CuCr_2O_4$ they are compressed.

16.7   Problems

16.1 Suppose the connection of tetrahedra shown in Fig. 116(a) were continued to form a layer. What would the composition be?

16.2 Why are the MX_3 strands shown in Fig. 94 compatible only with a hexagonal closest-packing of the X atoms?

16.3 What structure types would you expect for TiN, FeP, FeSb, CoS, and CoSb?

16.4 Why are CdI_2 and BiI_3 much more susceptible to stacking faults than $CaBr_2$ or RhF_3?

16.5 What fraction of the tetrahedral interstices are occupied in solid Cl_2O_7?

16.6 Decide whether the following compounds should form normal or inverse spinels using the ligand field stabilization energy as the criterion:
MgV_2O_4, VMg_2O_4, $NiGa_2O_4$, $ZnCr_2S_4$, $NiFe_2O_4$.

17 Physical Properties of Solids

The majority of the materials we use and handle every day are solid. We take advantage of their physical properties in manifold ways. The properties are intimately related to the structures. In the following we will deal only with a few properties that are directly connected with some structural aspects; many other properties such as electrical and thermal conductivity, optical transparency and reflectivity, color, etc. require the discussion of more sophisticated theories that are beyond the scope of this book.

17.1 Mechanical Properties

In addition to specific properties of interest for a particular application of a material, its elasticity, compressive and tensile strength, deformability, hardness, wear-resistance, brittleness, and cleavability also determine whether an application is possible. No matter how good the electric, magnetic, chemical or other properties are, a material is of no use if it does not fulfil mechanical requirements. These depend to a large extent on the structure and on the kind of chemical bonding. Mechanical properties usually are anisotropic, i.e. they depend on the direction of the applied force.

A three-dimensional network of strong covalent bonds as in diamond results in high hardness and compressive strength. It also accounts for a high tensile strength; in this case it is sufficient if the covalent bonds are present in the direction of the tensile stress. Hardness can be determined in a qualitative manner by the MOHS scratch test: a material capable of scratching another is the harder of the two. Standard materials on the MOHS scale are talc at the lower end (hardness 1) and diamond (hardness 10) at the upper end. Because its structure consists of electrically uncharged layers, talc is soft; only VAN DER WAALS forces act between the layers (cf. Fig. 107), and the layers can easily slide over each other. The same applies to graphite and MoS_2, which are used as lubricants. Crystals consisting of stiff parallel chain molecules have strong bonding forces in the chain direction and weak ones in perpendicular directions. They can be cleaved to form fiber bunches.

Ionic crystals have moderate to medium hardness, those incorporating highly charged ions being harder (e.g. NaCl hardness 2, CaF_2 hardness 4). Quartz with its network of polar covalent bonds is harder (hardness 7). The surfaces of materials with hardness below 7 become lusterless in everyday use because they continually suffer scratching from quartz particles in dust (which accounts for an advantage of glass over Plexiglass). The differing cohesion due to covalent and to ionic bonds in different directions is apparent in micas. Micas consist of anionic

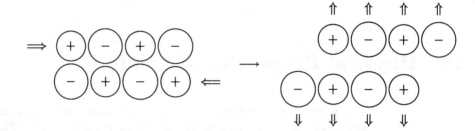

Fig. 131
Shearing forces
exerted on an
ionic crystal
(left) result in
cleavage (right)

layers and intercalated cations. Within the layers atoms are held together by (polar) covalent bonds. Micas can easily be cleaved parallel to the layers, which allows the manufacture of plates with sizes of several square decimeters and a thickness of less than 0.01 mm.

Ionic crystals can be cleaved in certain directions. Fig. 131 shows why the exertion of a force results in cleavage: if two parts of a crystal experience a mutual displacement by a shearing force, ions of like charges come to lie side by side and repel each other. The displacement is easiest along planes which have the fewest cation–anion contacts. In rock salt, for example, one encounters one Cl^- neighbor when looking from an Na^+ ion in the direction parallel to an edge of the unit cell, two Cl^- neighbors if one looks in the direction diagonal to a face, and three Cl^- neighbors in the direction of a body diagonal. An NaCl crystal is most easily cleaved perpendicular to a cell edge.

Metals behave differently since the metal atoms are embedded in an electron gas. The attractive forces remain active even after mutual displacement of parts of a crystal has occurred. Metals therefore can be deformed without fracture.

Most ceramic materials are oxides (MgO, Al_2O_3, ZrO_2, silicates), though some are nitrides (BN, Si_3N_4) or carbides (SiC). Because of the short range of action of the chemical bonds, the material suffers a substantial loss of strength once a rupture has begun. The resulting brittleness is one of the most severe drawbacks of ceramic materials. This problem has been largely solved for one material, zirconium dioxide. ZrO_2 adopts several modifications: at temperatures above 2370 °C it has the cubic CaF_2 structure (Zr atoms with c.n. 8), between 1170 and 2370 °C it has a slightly distorted, tetragonal CaF_2 structure (Zr coordination 4+4), and below 1170 °C baddeleyite is the stable form; this is a more distorted variety of the CaF_2 type in which a Zr atom only has the coordination number 7. By doping with a few percent of Y_2O_3 (or certain other oxides), the tetragonal form can be stabilized down to room temperature. Compared to the tetragonal modification, baddeleyite requires 7 % more volume. For this reason *pure* ZrO_2 is not appropriate as a high temperature ceramic: it cracks during heating when the transition temperature of 1170 °C is reached. But it is precisely this volume effect which is taken advantage of in order to reduce the brittleness, thus rendering ZrO_2 a high performance ceramic material. Such material consists of "partially stabilized" tetragonal ZrO_2, i.e. it is maintained metastable in this modification by doping. The mechanical forces have their maximum at the tip of a crack, and this is where the crack propagates in common materials. In metastable tetragonal ZrO_2, however, mechanical strain at the tip of a crack induces a transition to the baddeleyite form at this site and by the volume expansion the crack is sealed.

17.2 Piezoelectric and Ferroelectric Properties

The Piezoelectric Effect

Within a crystal, consider an atom with a positive partial charge that is surrounded tetrahedrally by atoms with negative partial charges. The center of gravity of the negative charges is at the center of the tetrahedron. By exerting pressure on the crystal in an appropriate direction, the tetrahedron will experience a distortion (Fig. 132) and the center of gravity of the negative charges will no longer coincide with the position of the positive central atom; an electric dipole has developed. If there are inversion centers in the structure, then for every tetrahedron there is another tetrahedron which has the exact opposite orientation; the electric fields of the dipoles compensate each other. If, however, all tetrahedra have the same orientation or some other mutual orientation that does not allow for a compensation, then the action of all dipoles adds up and the whole crystal becomes a dipole. Two opposite faces of the crystal develop opposite electric charges. Depending on the direction of the acting force, the faces being charged are either the two faces experiencing the pressure or two other faces in a perpendicular or an inclined direction.

This described *piezoelectric effect* is reversible. If the crystal is introduced into an external electric field, it experiences a contraction or an elongation. Crystals can only be piezoelectric if they are non-centrosymmetric. Sphalerite (ZnS), turmaline, ammonium chloride and quartz are examples. The effect is used in the quartz resonators that beat time in electronic watches and computers. The quartz resonator is a sheet cut from a quartz crystal in an appropriate

Fig. 132
Explanation of the piezoelectric effect: external pressure causes the deformation of a coordination tetrahedron, resulting in a shift of the centers of gravity of the electric charges

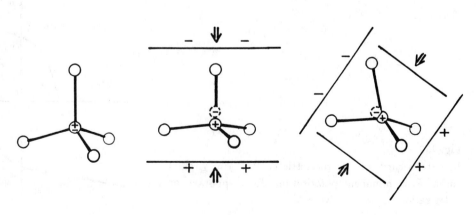

direction. Metal coatings act as electrical contacts. Mechanical vibrations are induced in the quartz with the aid of electric pulses; these vibrations have an exactly defined frequency and produce a corresponding alternating electric field. Piezoelectric crystals also serve whenever mechanical pulses are to be converted to electrical signals or vice versa, e.g. in microphones or in the production of ultrasound.

Ferroelectricity

In some crystalline substances the centers of gravity of positive and negative charges do not coincide in the first place, i.e. permanent dipoles are present. Concerning the electrical properties, the following cases can be distinguished.

If the crystal is centrosymmetric (or, generally, nonpolar), the action of the dipoles is mutually compensated, and no special effect can be observed.

A *paraelectric* substance is not polarized macroscopically because the dipoles are oriented randomly. However, they can be oriented by an external electric field (orientation polarization). The orientation is counteracted by thermal motion, i.e. the degree of polarization decreases with increasing temperature.

An *electret* is a crystal which has dipoles oriented permanently in one direction. The crystal therefore is a macroscopic dipole.

In a *ferroelectric* substance all dipoles are also oriented uniformly, but only within one *domain*. The orientation differs from domain to domain. As a whole, the dipole moments of different domains compensate each other in a "virginal" sample. If an increasing external electric field acts on the sample, those domains whose polarization corresponds to the direction of the electric field will grow at the expense of the remaining domains. The total polarization of the crystal increases (curve j in Fig. 133). Finally, if the external field is strong enough, the whole crystal is one large domain, and the polarization continues to increase only slightly with increasing electric field (curve s; the continuing increase is due to the normal dielectric polarization which takes place in any substance

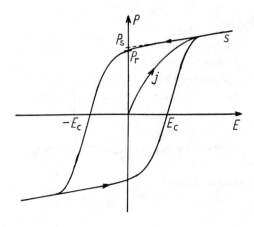

Fig. 133
Hysteresis curve of a ferroelectric crystal. j = virginal curve, P_r = remanent polarisation, P_s = spontaneous polarization, E_c = coercive field

by polarization of the electrons). If the external electric field is removed, a *remanent polarization* P_r remains, i.e. the crystal now is an electret. In order to remove the remanent polarization, an oppositely oriented electric field with the field E_c has to be applied; this is the coercive field. The value of P_s, the spontaneous polarization, corresponds to the polarization within a domain.

Above a specific temperature, the CURIE temperature, a ferroelectric substance becomes paraelectric since the thermal vibrations counteract the orientation of the dipoles. The coordinated orientation of the dipoles taking place during the ferroelectric polarization is a *cooperative phenomenon*. This behavior is similar to that of ferromagnetic substances, which is the reason for its name; the effect has to do nothing with iron (it is also called seignette or rochelle electricity).

The polarization induced by the electric field is considerably larger than in a nonferroelectric substance, and therefore the dielectric constants are much larger. $BaTiO_3$ in particular has practical applications in the manufacture of capacitors with large capacitance. Further examples include SbSI, KH_2PO_4, and $NaNO_2$, as well as certain substances which have a distorted perovskite structure such as $LiNbO_3$ and $KNbO_3$. Fig. 134 shows how all nitrite ions in sodium nitrite are oriented in one direction below 164 °C, thus producing a macroscopic dipole moment. It also shows how the differently oriented domains alternate as long as there has been no electric field to shift all the NO_2^- ions into the same orientation. Above the CURIE temperature of 164 °C all NO_2^- ions are randomly oriented and $NaNO_2$ is paraelectric. In sodium nitrite the ferroelectric polarization only occurs in one direction. In $BaTiO_3$ it is not re-

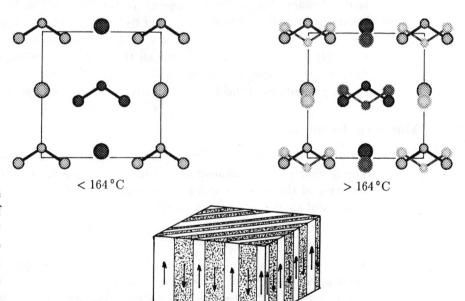

$< 164\,°C$ $> 164\,°C$

Fig. 134
Structure of
$NaNO_2$ below
and above the
CURIE point.
Bottom: domains
in a ferroelectric
crystal of $NaNO_2$

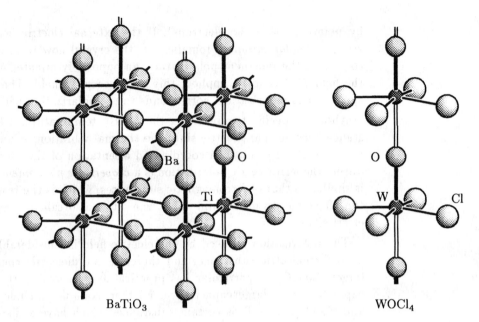

Fig. 135

Structure of
ferroelectric
BaTiO$_3$ and the
analogous
situation in the
electret WOCl$_4$

BaTiO$_3$ WOCl$_4$

stricted to one direction. BaTiO$_3$ has the structure of a distorted perovskite
between 5 and 120 °C. Due to the size of the Ba^{2+} ions, which form a closest
packing of spheres together with the oxygen atoms, the octahedral interstices
are rather too large for the titanium atoms, and these consequently do not
occupy exactly the octahedron centers. The titanium atom in an octahedron
is shifted towards one of the O atoms, the direction of the shifting being the
same for all octahedra of one domain (Fig. 135). The result is a polarization
in the domain. The shifting is similar for the W atoms in WOCl$_4$, which has
square-pyramidal molecules associated to form a strand with alternating short
and long W–O distances. Above the CURIE temperature of 120 °C BaTiO$_3$ has
the cubic perovskite structure, with all titanium atoms occupying the octahedron centers (averaged over time). A considerably higher CURIE temperature
(1470 °C) and also a much larger polarization has been found for LiNbO$_3$.

17.3   Magnetic Properties

An unpaired electron, taken as a particle, executes a spin about its own axis.
The mechanical spin momentum is related to a spin vector which specifies the
direction of the rotation axis and the magnitude of the momentum. The spin
vector **s** of an electron has an exactly defined magnitude:

$$|\mathbf{s}| = \frac{h}{2\pi}\sqrt{s(s+1)} = \frac{h}{4\pi}\sqrt{3}$$

The spin quantum number s is used to characterize the spin. It can have only
the one numerical value of $s = \frac{1}{2}$. $h = 6.6262 \cdot 10^{-34}$ Js = PLANCK's constant.
   The spin is associated with a magnetic moment, i.e. an electron behaves
like a tiny bar magnet. An external magnetic field exerts a force on an electron,

resulting in a precession of the electron about the direction of the magnetic field which is similar to the precession of a top; the rotation axis of the electron thus is inclined relative to the magnetic field. Quantum theory only permits two values for the inclination; they are expressed by the magnetic spin quantum number m_s, which has the value of $m_s = +s = +\frac{1}{2}$ or $m_s = -s = -\frac{1}{2}$. The two inclinations are also called "parallel" and "antiparallel", although the spin vectors are not really exactly parallel or antiparallel to the magnetic field.

Two electrons in an atom exert an influence on each other, i.e. their spins are coupled. Two electrons are termed paired if they coincide in all of their quantum numbers except the spin quantum number. In such an electron pair the magnetic moments of the electrons compensate each other. Unpaired electrons in different orbitals tend to orient their spins in the same direction ("parallel") and thus produce an accordingly larger magnetic field (HUND's rule); they have the same magnetic spin quantum number and differ in some other quantum number.

Substances having only paired electrons are *diamagnetic*. When they are introduced into an external magnetic field, a force acts on the electrons, i.e. an electric current is induced; the magentic field of this current is opposed to the external field (LENZ's rule). As a result, the substance experiences a repulsion by the external magnetic field; the repulsive force is rather weak.

In a *paramagnetic* substance unpaired electrons are present. Frequently the unpaired electrons can be assigned to certain atoms.[*] When an external magnetic field acts on a paramagnetic substance, the magnetic moments of the electrons adopt the orientation of this field, the sample is magnetized and a force pulls the substance into the field. The magnetization can be determined quantitatively by measuring this force. Thermal motion of the atoms prevents a complete orientation, and higher temperatures therefore cause a smaller degree of magnetization.

The additional magnetic field produced by the orientation is a measure of the magnetization M. It is proportional to the external magnetic field H:

$$M = \chi H$$

The proportionality factor χ is the *magnetic susceptibility*. The magnetization and consequently also the susceptibility depend on the number of orientable particles in a given volume. A volume independent, material specific magnitude is the *molar susceptibility* χ_{mol}:

$$\chi_{\text{mol}} = \chi V_{\text{mol}} = \chi_{\text{g}} M_{\text{mol}}$$

V_{mol} is the molar volume, M_{mol} the molar mass and χ_{g} the commonly specified gram-susceptibility (referred to 1 g of the sample).

Taking the susceptibility, we can classify materials according to their magnetic properties in the following way:

---

[*]An unpaired electron cannot always be assigned to a specific atom. E.g., in NO_2 the unpaired electron occupies a "delocalized" molecular orbital and simultaneously belongs to all three atoms.

$\chi_{mol} < 0$ diamagnetic

$\chi_{mol} > 0$ paramagnetic

$\chi_{mol} \gg 0$ ferromagnetic

Paramagnetism

The temperature dependence of the molar susceptibility of a paramagnetic substance follows the CURIE–WEISS law:

$$\chi_{mol} = \frac{C}{T - \Theta} \tag{28}$$

where T = absolute temperature, C = CURIE constant, and Θ = WEISS constant. For $\Theta = 0$ the equation is simplified to the classic CURIE law $\chi_{mol} = C/T$.

The following discussion is restricted to the case of a substance containing only one species of paramagnetic atoms (atoms with unpaired electrons). The *magnetic moment* μ is used to specify how magnetic an atom is. An increasing magnetic moment results in an increasing susceptibility; the quantitative relation is given by means of the CURIE constant:

$$C = \frac{(N_A \mu)^2}{3R} \tag{29}$$

N_A = AVOGADRO's number, R = gas constant.

The magnetic moment of an isolated electron has a specific value of

$$\mu_s = \frac{eh}{2\pi m_e}\sqrt{s(s+1)} = 2\mu_B\sqrt{3} \tag{30}$$

$\mu_B = eh/(4\pi m_e) = 9.274 \cdot 10^{-24}$ JT^{-1} is termed the "BOHR magneton".[*]

The coupling of the spins of the electrons in an atom is accounted for by adding their magnetic spin quantum numbers. Since they add up to zero for paired electrons, it is sufficient to consider only the unpaired electrons. The spins of n unpaired electrons add up according to HUND's rule to a total spin quantum number $S = \frac{1}{2}n$. The magnetic moment of these n electrons, however, is not the scalar sum of the single magnetic moments; the spin momenta must be added vectorially, taking into account the particular directions they may adopt according to quantum theory. The addition of the spin momenta yields a total spin momentum with the magintude:

$$|S| = \frac{h}{2\pi}\sqrt{S(S+1)}$$

which is related to a magnetic moment

$$\mu = 2\mu_B\sqrt{S(S+1)} \tag{31}$$

Up to now we have only considered the magnetism due to the electron spin. An electron orbiting in an atom is a circular electric current that also is surrounded by a magnetic field. For the corresponding orbit magnetism similar

---

[*]e = unit charge, h = Planck constant, m_e = electron mass; 1 tesla is the unit of the magnetic flux density, 1 T = 1 Vsm^{-2}

considerations apply as for the spin magnetism; in addition, matters are complicated by a coupling between spin and orbit magnetism. We will not consider the consequences arising from this in detail since the magnetic properties of a solid depend mainly on the spin magnetism. This is because almost no reorientation of the electron orbits takes place in an external magnetic field since the atoms and the chemical bonds in a solid are spatially fixed. The spatial fixation concerns the electron orbits of the valence electrons, including the nonbonding electrons in the valence shell, which particularly also includes the *d* electrons of transition metal complexes. Especially for compounds of the 3*d* metals the observed magnetism usually agrees quite well with the calculated "spin-only" value. In this case according to equations (28), (29) and (31) we obtain:

$$\chi_{mol} = 4\frac{(N_A \mu_B)^2}{3R(T - \Theta)} S(S + 1) \tag{32}$$

and

$$\mu_{eff} = \frac{\mu}{\mu_B} = \sqrt{0.8\frac{T^2 mol}{J\,K}\,\chi_{mol}(T - \Theta)} = 2\sqrt{S(S + 1)} \tag{33}$$

because the product of the constants happens to be almost exactly 0.800 $T^2 mol\,J^{-1}K^{-1}$.

If the measured magnetic susceptibility dependent on temperature agrees with equation (32) or (33), then the substance is paramagnetic and the computed value of S yields information about the number of unpaired electrons.

Especially for compounds of the heavier transition metals (4*d* and 5*d* metals) the spin-only value shows less exact agreement with the actual magnetism. In this case the orbit magnetism is also of minor importance, but the coupling of the magnetic moments of spin and orbit often can not be neglected and HUND's rule usually can no longer be applied. The spin–orbit coupling depends on the electronic configuration; frequently, for d^1 to d^3 it causes a reduced magnetic moment, and for d^5 to d^9 an increased magnetic moment. For more details the specialist literature should be consulted [50,51].

The 4*f* electrons of lanthanoid ions, being shielded by the fully occupied, spherical shells 5*s* and 5*p*, experience almost no influence from the ligands; their orbits can orient themselves in a magnetic field, so that in this case the orbit magnetism has to be considered; however, we do not discuss this aspect here.

Ferro-, Ferri- and Antiferromagnetism

The term ferromagnetism reflects the fact that iron shows this effect, but it is by no means restricted to iron or iron compounds. Ferromagnetism is a *cooperative phenomenon*, i.e. many particles in a solid behave in a coupled manner. Paramagnetic atoms or ions exert influence on each other over extended regions.

In a ferromagnetic substance the magnetic moments of adjacent atoms orient themselves mutually parallel. However, a material that has not been treated magnetically exhibits no macroscopic magnetic moment. This is the case only when all magnetic moments have been oriented parallel with the aid of an external magnetic field, so that their action is added up. Permanent magnets retain their magnetization even after the external field has been removed. The lack of a macroscopic magnetic moment in a previously untreated sample is due to the presence of numerous domains (WEISS domains). In each domain the orientation of all spins is parallel, but it differs from domain to domain. An external magnetic field causes the growth of those domains oriented parallel with the magnetic field at the expense of the other domains. When the spins of all domains have been oriented, saturation has taken place. To achieve this state, a magnetic field with some minimum field strength is required; its magnitude is dependent on the material.

A hysteresis curve shows this kind of behavior; it is like the hysteresis of a ferroelectric material (Fig. 136). Starting with an untreated sample, an increasing magnetic field causes an increasing magnetization until saturation has been reached. After turning off the magnetic field, there is some loss of magnetization, but a remanent magnetization R is retained. By reversing the magnetic field, the spins experience a reorientation. The minimum magnetic field required for this is the coercive force or coercive field K. Depending on the application, magnetic materials with different magnetic "hardness" are required. For example, a permanent magnet in an electrical motor must have a high coercive force in order to maintain its magnetization. A magnetic pigment on a diskette, however, should have a medium coercive force, so that it can be remagnetized quickly, but, on the other hand, it is not supposed to lose the stored information. Small coercive forces are required whenever frequent and fast remagnetizations are required, as for example in recording heads.

Above a critical temperature T_C, the CURIE temperature, a ferromagnetic

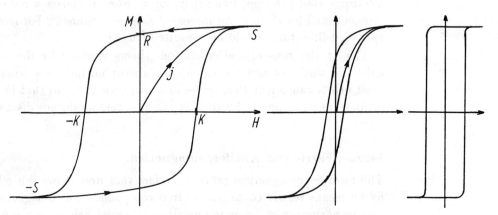

Fig. 136

Hysteresis curves for three ferromagnetic materials. Left, for a magnetically "hard" material; center, for a magnetically "soft" material; right, for a material appropriate for data storage, which has a rectangular hysteresis. j = virginal curve, S = saturation magnetization, R = remanent magnetization, K = coercive force

Table 24 Coupling of the spin vectors related to cooperative magnetic effects

	spin orientation within a domain	example
ferromagnetism	↑↑	EuO (NaCl type)
antiferromagnetism	↑↓	MnF_2 (rutile type)
ferrimagnetism		
NÉEL type	↑↓	$NiFe_2O_4$ (inverse spinel)
YAFET-KITTEL type	↑↘	$MnCr_2S_4$ (spinel)
helix structure	↑↗↗ → ↘↓↓ …	$MnCr_2O_4$ (spinel)

material becomes paramagnetic, since thermal motion inhibits the orientation of the WEISS domains.

The coupling of the magnetic moments of the atoms can also result in spins having opposite orientations; in this case the material is *antiferromagnetic*. At very low temperatures the total magnetic moment is zero. With increasing temperatures the thermal motion interferes with the antiparallel orientation of all particles and the magnetic susceptibilty increases; at even higher temperatures, thermal motion increasingly causes a random distribution of the spin orientations and the magnetic susceptibility decreases again, as in a paramagnetic material. Therefore, an antiferromagnetic material exhibits a maximum susceptibility at a certain temperature, the NÉEL temperature.

In a *ferrimagnetic* material the situation is the same as in an antiferromagnetic material, but the particles bearing opposite magnetic moments occur in different quantities and/or their magnetic moments differ in magnitude. As a consequence, they do not compensate each other even at very low temperatures. The behaviour in a magentic field is like that of ferromagnetic materials. The spins can also be coupled to adopt more than two different orientations (Table 24). The order occurring among the spins of the atoms in the unit cell can be determined experimentally by neutron diffraction. Since a neutron itself has a spin and a magnetic moment, it is diffracted by an atom to an extent which depends on the orientation of the magnetic moment.

What determines the way the spins couple? Parallel orientation always occurs when the corresponding atoms act directly on one another. This is the case in pure metals like iron or nickel, but also in EuO (NaCl type). Antiparallel orientation usually occurs when two paramagnetic particles interact indirectly by means of the electrons of an intermediate particle which itself is not paramagnetic ("super exchange"). This is the case in the commercially important spinels and garnets.

In $NiFe_2O_4$, an inverse spinel $Fe_T^{3+}[Ni^{2+}Fe^{3+}]_O O_4$, the spins of the octahedral sites are parallel with one another; the same applies for the tetrahedral sites. The interaction between the two kinds of sites is mediated by super exchange via the oxygen atoms. High-spin states being involved, Fe^{3+} (d^5) has 5 unpaired electrons, and Ni^{2+} (d^8) has 2 unpaired electrons. The coupled parallel spins at the octahedral sites add up to a spin of $S = \frac{7}{2}$; it is opposed to the spin of $S = \frac{5}{2}$ of the Fe^{3+} particles at the tetrahedral sites. A total spin of $S = 1$ remains which is equivalent to two unpaired electrons.

Magnetic Materials of Practical Importance

Iron is a material whose ferromagnetic properties have been applied for a long time. It becomes paramagnetic when heated above the CURIE temperature of 766 °C; this does not involve a phase transition of the body-centered cubic structure (a phase transition to cubic closest-packing occurs at 906 °C). Iron is applied in electric motors and in transformers. Finely dispersed iron powder ("iron pigment") serves as magnetic material in data storage devices (since it is pyrophoric it is stabilized by plating it by evaporation with a Co–Cr alloy). A disadvantage of iron is its electrical conductivity; an alternating magnetic field induces electric eddy currents which cause energy losses by heating the iron. Using stacks of mutually insulated iron plates can decrease but not fully avoid the eddy currents. Wherever there are no alternating magnetic fields, the electric conductivity causes no trouble, for example in certain applications of permanent magnets. Permanent magnets with especially high coercivity are made of $SmCo_5$.

Nearly no eddy current losses occur in electrically insulating magnetic materials. This is one of the reasons for the importance of oxidic materials, especially of spinels and garnets. Another reason is the large variability of the magnetic properties that can be achieved with spinels and garnets of different compositions. The tolerance of the spinel structure towards substitution at the metal atom sites and the interplay between normal and inverse spinels allow the adaptation of the properties to given requirements.

Spinel ferrites are iron-containing spinels $M^{II}Fe_2O_4$. They are magnetically "soft" to "medium hard". The medium hard ferrites serve in form of pigments as data storage materials, especially γ-Fe_2O_3 (diskettes, recording tapes), and γ-Fe_2O_3 with additives of $CoFe_2O_4$ (video cassettes). γ-Fe_2O_3 is a spinel with defect structure, $Fe_T^{III}[Fe_{1.67}^{III}\square_{0.33}]_O O_4$. Fe_3O_4 is applied in "magnetic liquids" that are used to seal bearings against vacuum. These are suspensions of magnetic pigments in oil; in a magnetic field the pigment collects in the region with the most intense field and causes an increase in the density and viscosity of the liquid. In high frequency electronics (e.g. for deflection magnets in television tubes) manganese-zinc ferrites are used at frequencies up to 500 kHz, and nickel-zinc ferrites up to 200 MHz. At even higher frequencies (microwaves, radar; 500 MHz to 300 GHz) garnet-ferrites are applied. Garnet is an orthosilicate, $Mg_3Al_2(SiO_4)_3$, which has a complicated cubic structure. The structure is retained when all metal atoms are trivalent in the sense of the following substitution:

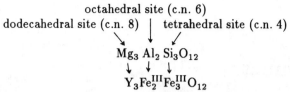

$$\text{Mg}_3 \text{ Al}_2 \text{ Si}_3 \text{O}_{12}$$
$$\downarrow \quad \downarrow \quad \downarrow$$
$$\text{Y}_3 \text{Fe}_2^{III} \text{Fe}_3^{III} \text{O}_{12}$$

In yttrium iron garnet $\text{Y}_3\text{Fe}_5\text{O}_{12}$ ("YIG") a ferrimagnetic coupling (super exchange) is active between the octahedral and the tetrahedral sites. Since the tetrahedral sites are in excess, the magnetic moments do not compensate each other.

Hexagonal ferrites of the magnetoplumbite type serve as magnetic "hard" materials. They have high coercivities and are used as nonconducting permanent magnets, for example in electric motors, generators, and closet locks. Structurally, they are related to spinels, but with some of the oxygen atoms substituted by larger cations like Ba^{2+} or Pb^{2+}. The two main types are: $\text{BaFe}_{12}\text{O}_{19}$ ("M phase") and $\text{Ba}_2\text{Zn}_2\text{Fe}_{12}\text{O}_{22}$ ("Y phase"). $\text{BaFe}_{12}\text{O}_{19}$ is gaining in importance for its use in high capacity diskettes.

18 Symmetry

A characteristic feature of any crystal is its symmetry. It not only serves to describe important aspects of a structure, but is also related to essential properties of a solid. For instance, quartz crystals could not exhibit the piezoelectric effect if quartz did not have the appropriate symmetry; this effect is the basis for the application of quartz in watches and electronic devices. Knowledge of the crystal symmetry is also of fundamental importance in crystal structure analysis.

In order to designate symmetry in a compact form, symmetry symbols have been developed. Two kinds of symbols are used: the *Schoenflies symbols* and the *Hermann–Mauguin symbols*, which are also called *international symbols*. Historically, Schoenflies symbols were developed first; they continue to be used in spectroscopy and to designate the symmetry of molecules. However, since they are less appropriate for describing the symmetry in crystals, they are now scarcely used in crystallography. We therefore discuss primarily the Hermann–Mauguin symbols. In addition, there are graphical symbols which are used in figures.

18.1 Symmetry Elements and Symmetry Operations

A symmetry operation is the motion of an object into a position that cannot be distinguished from its original position. The operation can be repeated infinitely many times. A symmetry element serves to characterize the symmetry operation. The symmetry operations are the following:

1. Translation. Shift in a specified direction by a specified length. The corresponding symmetry element is the *translation vector*. For example:

Crystals possess translation symmetry in three dimensions. It is characterized by three non-coplanar translation vectors $\mathbf{a}$, $\mathbf{b}$ and $\mathbf{c}$. These are identical to the three base vectors used to define the unit cell (section 2.2). Any translation vector $\mathbf{r}$ in the crystal can be expressed as the vectorial sum of the three base vectors, $\mathbf{r} = u\mathbf{a} + v\mathbf{b} + w\mathbf{c}$, where u, v and w are integers. Translation symmetry is the most important symmetry property of a crystal. In the Hermann–Mauguin symbol the three-dimensional translation symmetry is expressed by a capital letter which also allows the distinction of primitive and centered crystal lattices (cf. Fig. 6, p. 9):

P = primitive
A, B or C = base-centered in the **bc**-, **ac** or **ab** plane, respectively
F = face-centered (all faces)
I = body-centered*
R = rhombohedral

2. Rotation about some axis by an angle of $360/N$ degrees. The symmetry element is an N-fold *rotation axis*. The multiplicity N has to be an integer, otherwise the condition of infinite repeatability of the symmetry operation would not be fulfilled. Every object possesses infinitely many axes with $N = 1$, since an arbitrary rotation by $360°$ returns the object into its original position. The symbol for the onefold rotation axis is used for objects that have no symmetry other than translation symmetry. The Hermann–Mauguin symbol for an N-fold rotation axis is the number N; the Schoenflies symbol is C_N (cf. Fig. 137):

	Hermann–Mauguin symbol	Schoenflies symbol	graphical symbol	
onefold rotation axis	1	C_1	none	
twofold rotation axis	2	C_2	●	axis perpendicular to the plane of the paper
			← →	axis parallel to the plane of the paper
threefold rotation axis	3	C_3	▲	
fourfold rotation axis	4	C_4	◆	
sixfold rotation axis	6	C_6	⬢	

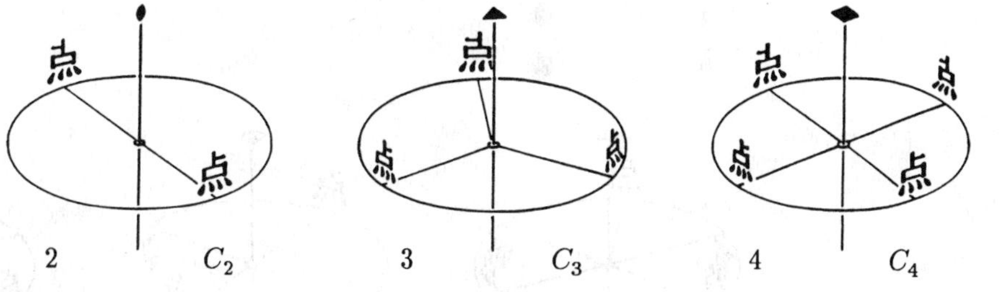

Fig. 137
Examples for rotation axes.[†] In each case the Hermann–Mauguin symbol is given on the left side, and the Schoenflies symbol on the right side

---

innenzentriert in German.

[†] 点 = chinese symbol for point. Reproduction of the figures with permission by G. Thieme Verlag from the book *Schwingungsspektroskopie* by J. Weidlein, U. Müller and K. Dehnicke.

3. Reflection. The symmetry element is a *reflection plane* (Fig. 138).
Hermann–Mauguin symbol: m.
Graphical symbols:

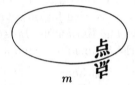

reflection plane perpendicular   reflection plane parallel
to the plane of the paper   to the plane of the paper

4. Inversion. "Reflection" through a point (Fig. 138). This point is the
symmetry element and is called *inversion center* or *center of symmetry*.
Hermann–Mauguin symbol: $\bar{1}$ ("one bar"). Schoenflies symbol: i.
Graphical symbol: o

Fig. 138
Examples of a
reflection plane
and of an
inversion center

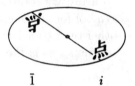

m $\bar{1}$ i

5. Rotoinversion. The symmetry element is a *rotoinversion axis* or, for
short, an *inversion axis*. This refers to a *coupled* symmetry operation which
involves two motions: take a rotation through an angle of $360/N$ degrees immediately followed by an inversion at a point located on the axis (Fig. 139):

Hermann–   graphical
Mauguin   symbol
symbol

$\bar{1}$ o   identical with an inversion center

$\bar{2} = m$ identical with a reflection plane perpendicular to the axis

$\bar{3}$ ▲

$\bar{4}$ ◆

$\bar{6}$ ⬢

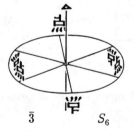 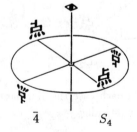 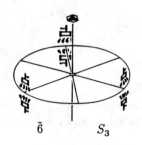

$\bar{3}$ S_6 $\bar{4}$ S_4 $\bar{6}$ S_3

Fig. 139
Examples for inversion axes. If they are considered to be rotoreflection axes, they have the multiplicities
expressed by the Schoenflies symbols S_N

If N is an even number, the inversion axis automatically contains a rotation axis with half the multiplicity. If N is an odd number, automatically an inversion center is present. This is expressed by the graphical symbols. If N is even but not divisible by 4, automatically a reflection plane perpendicular to the axis is present.

A **rotoreflection axis** is a coupled symmetry operation of a rotation and a reflection at a plane perpendicular to the axis. Rotoreflection axes are identical to inversion axes, but the multiplicities do not coincide in all cases (Fig. 139). In the Hermann–Mauguin notation only inversion axes are used, and in the Schoenflies notation only rotoreflection axes are used, the symbol for the latter being S_N.

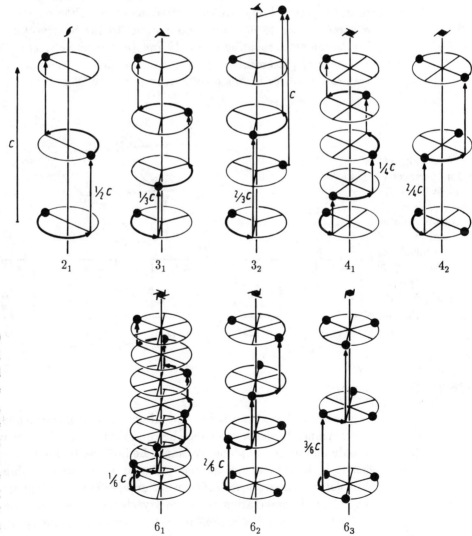

Fig. 140 Screw axes and their graphical symbols. The axes $3_1, 4_1, 6_1,$ and 6_2 are right-handed; $3_2, 4_3, 6_5,$ and 6_4 are the corresponding left-handed screw axes

6. Screw rotation. The symmetry element is a *screw axis*. It can only occur if there is translation symmetry in the direction of the axis. The screw axis results when a rotation of $360/N$ degrees is coupled with a displacement parallel to the axis. The Hermann–Mauguin symbol is N_M ("N sub M"); N expresses the rotational component and the fraction M/N is the displacement component as a fraction of the translation vector. Screw axes can be right or left-handed. Screw axes that can occur in crystals are shown in Fig. 140. Single polymer molecules can also have non-crystallographic screw axes, e.g. 10_3 in polymeric sulfur.

7. Glide reflection. The symmetry element is a *glide plane*; it can only occur if translation symmetry is present parallel to the plane. At the plane, reflections are performed, but every reflection is coupled with an immediate displacement parallel to the plane. The Hermann–Mauguin symbol is a, b, c, n or d, the letter designating the direction of the glide referred to the unit cell. a, b and c refer to displacements parallel to the base vectors **a**, **b** and **c**, the displacements amounting to $\frac{1}{2}a$, $\frac{1}{2}b$ and $\frac{1}{2}c$, respectively. The glide planes n and d involve displacements in a diagonal direction by amounts of $\frac{1}{2}$ and $\frac{1}{4}$ of the translation vector in this direction, respectively (Fig. 141).

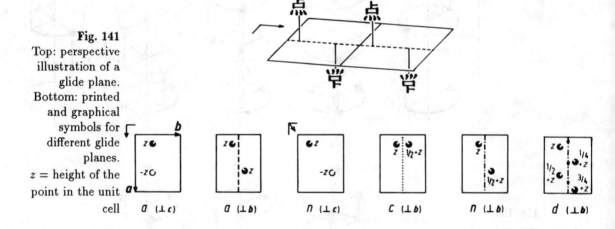

Fig. 141
Top: perspective illustration of a glide plane.
Bottom: printed and graphical symbols for different glide planes.
z = height of the point in the unit cell

18.2 Point Groups

A geometric object can possess several symmetry elements simultaneously. However, symmetry elements cannot be combined arbitrarily. For instance, if there is only one reflection plane, it cannot be inclined to a symmetry axis (the axis has to be in the plane or perpendicular to it). Possible combinations *excluding translation symmetry* are called *point groups*. This term expresses the fact that any allowed combination of symmetry elements has one unique point or one unique axis which is common to all the symmetry elements.

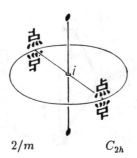

Fig. 142
The combination of a twofold rotation
axis and a reflection plane perpendicular to it automatically results in an
inversion center

$2/m$ C_{2h}

When two symmetry elements are combined, a third symmetry element can
result automatically. For example, the combination of a twofold rotation axis
and a reflection plane perpendicular to it automatically results in an inversion
center at the site where the axis crosses the plane. It makes no difference which
two of the three symmetry elements are combined (2, m or $\bar{1}$), the third one
always results (Fig. 142).

Hermann–Mauguin Point Group Symbols

A point group symbol consists of a listing of the symmetry elements that are
present according to certain rules in such a way that their relative orientations
can also be recognized. In the *full Hermann–Mauguin symbol* all symmetry elements, with few exceptions, are listed. However, because they are more compact,
usually only the *short Hermann–Mauguin symbols* are cited; in these, symmetry axes that result automatically from mentioned symmetry planes are not
expressed; symmetry planes which are present are not omitted.

The following **rules** apply:

1. The orientation of symmetry elements is referred to a coordinate system xyz.
   If one symmetry axis is distinguished from the others by a higher multiplicity
   ("principal axis") or when there is only one symmetry axis, it is set as the z
   axis.

2. An inversion center is mentioned only if it is the only symmetry element
   present. The symbol then is $\bar{1}$. In other cases the presence or absence of an
   inversion center can be recognized as follows: it is present and only present if
   there is either an inversion axis with odd multiplicity ($\bar{N}$, with $N = 2n + 1$)
   or a rotation axis with even multiplicity and a reflection plane perpendicular
   to it (N/m, with $N = 2n$).

3. A symmetry element occurring repeatedly because it is multiplied by another
   symmetry element is mentioned only once.

4. A reflection plane that is perpendicular to a symmetry axis is designated by a
   fraction bar, e.g. $\dfrac{2}{m}$ or $2/m$ ("two over m") = reflection plane perpendicular
   to a twofold axis. However, reflection planes perpendicular to rotation axes
   with odd multiplicities are not usually designated in the form $3/m$, but as
   inversion axes like $\bar{6}$; $3/m$ and $\bar{6}$ express identical facts.

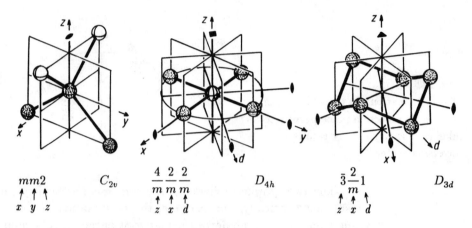

Fig. 143
Examples of three
point groups. The
letters under the
Hermann–
Mauguin symbols
indicate to which
directions the
symmetry
elements refer

5. The mutual orientation of different symmetry elements is expressed by the sequence in which they are listed. The orientation refers to the coordinate system. If the symmetry axis of highest multiplicity is twofold, the sequence is x–y–z, i.e. the symmetry element in the x direction is mentioned first etc.; the direction of reference for a reflection plane is *normal* to the plane. If there is an axis with a higher multiplicity, it is mentioned first; since it coincides by convention with the z axis, the sequence is different, namely z–x–d. The symmetry element oriented in the x direction occurs repeatedly because it is being multiplied by the higher multiplicity of the z axis; the bisecting direction between x and its next symmetry-equivalent direction is the direction indicated by d. See the examples in Fig. 143.

6. Cubic point groups have four threefold axes (3 or $\bar{3}$) that mutually intersect at angles of 109.47°. They correspond to the four body diagonals of a cube (directions **x+y+z**, **x–y+z**, **–x+y+z** and **x+y–z**, added vectorially). In the directions **x**, **y**, and **z** there are axes 4, $\bar{4}$ or 2, and there can be reflection planes perpendicular to them. In the six directions **x+y**, **x–y**, **x+z**, ... twofold axes and reflection planes may be present. The sequence of the reference directions in the Hermann–Mauguin symbol is **z, x+y+z, x+y**. The occurrence of a 3 in the *second position* of the symbol (direction **x+y+z**) gives evidence of a cubic point group. See Fig. 144.

Figs 144 and 145 list point group symbols and illustrate them by geometric figures. In addition to the short Hermann–Mauguin symbols the Schoenflies symbols are also listed. Full Hermann–Mauguin symbols for some point groups are:

	short	full		short	full
	mmm	$2/m\,2/m\,2/m$		$\bar{3}m$	$\bar{3}\,2/m\,1$
	$4/mmm$	$4/m\,2/m\,2/m$		$m\bar{3}m$	$4/m\,\bar{3}\,2/m$
	$6/mmm$	$6/m\,2/m\,2/m$		$m\bar{3}$	$2/m\,\bar{3}$

Schoenflies Point Group Symbols

The coordinate system of reference is taken with vertical principal axis (z axis). Schoenflies symbols are rather compact — they designate only a minimum of the

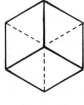 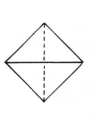 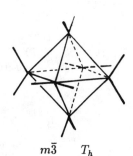

Fig. 144
Examples for
three cubic point
groups

$m\bar{3}m$ O_h $m\bar{3}m$ O_h $\bar{4}3m$ T_d $m\bar{3}$ T_h

symmetry elements present in the following way (the corresponding Hermann–Mauguin symbols are given in parentheses):

C_i = an inversion center is the only symmetry element ($\bar{1}$).

C_s = a reflection plane is the only symmetry element (m).

C_N = an N-fold rotation axis is the only symmetry element (N).

C_{Ni} (N odd) = an N-fold rotation axis and an inversion center ($\bar{N}$).

D_N = perpendicular to an N-fold rotation axis there are N twofold rotation axes ($N2$ if the value of N is odd; $N22$ if N is even).

C_{Nh} = there is one N-fold (vertical) rotation axis and one horizontal reflection plane (N/m).

C_{Nv} = an N-fold (vertical) rotation axis is at the intersection line of N vertical reflection planes (Nm if the value of N is odd; Nmm if N is even).

D_{Nh} = in addition to an N-fold (vertical) rotation axis there are N horizontal twofold axes, N vertical reflection planes and one horizontal reflection plane ($\bar{N}2/m$ if N is odd; $N/m2/m2/m$, for short N/mmm, if N is even).

D_{Nd} = the N-fold vertical rotation axis contains a $2N$-fold rotoreflection axis, N horizontal twofold rotation axes are situated at bisecting angles between N vertical reflection planes ($\bar{M}2m$ with $M = 2 \times N$).

S_N = there is only an N-fold (vertical) rotoreflection axis (cf. Fig. 139).

T_d = symmetry of a tetrahedron ($\bar{4}3m$).

O_h = symmetry of an octahedron and of a cube ($4/m\bar{3}2/m$, short $m\bar{3}m$).

T_h = symmetry of an octahedron without fourfold axes ($2/m\bar{3}$).

I_h = symmetry of an icosahedron and of a pentagonal dodecahedron ($\bar{5}32/m$, short $\bar{5}\bar{3}m$).

O, T and I = as O_h, T_h and I_h, but with no reflection planes (423, 23 and 532, respectively).

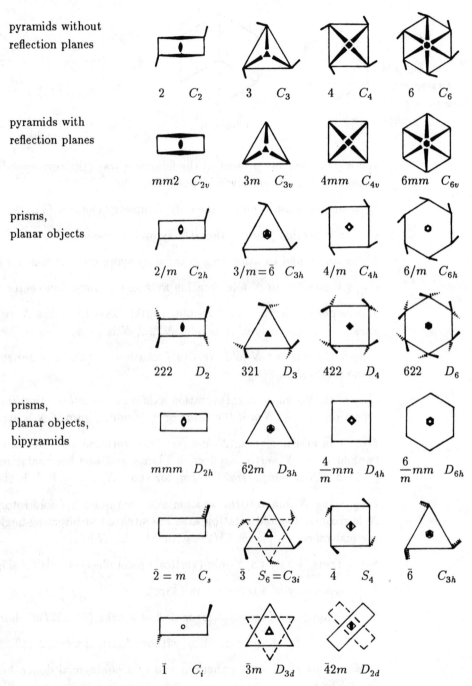

pyramids without
reflection planes

    2   C_2 3 C_3 4 C_4 6 C_6

pyramids with
reflection planes

  $mm2$ C_{2v} $3m$ C_{3v} $4mm$ C_{4v} $6mm$ C_{6v}

prisms,
planar objects

  $2/m$ C_{2h} $3/m = \bar{6}$ C_{3h} $4/m$ C_{4h} $6/m$ C_{6h}

  222   D_2 321 D_3 422 D_4 622 D_6

prisms,
planar objects,
bipyramids

  mmm D_{2h} $\bar{6}2m$ D_{3h} $\dfrac{4}{m}mm$ D_{4h} $\dfrac{6}{m}mm$ D_{6h}

  $\bar{2} = m$ C_s $\bar{3}$ $S_6 = C_{3i}$ $\bar{4}$ S_4 $\bar{6}$ C_{3h}

  $\bar{1}$ C_i $\bar{3}m$ D_{3d} $\bar{4}2m$ D_{2d}

Fig. 145

Symmetrical geometric figures and their point group symbols; in each case, the short Hermann–Mauguin
symbol is given to the left, and the Schoenflies symbol to the right

18.3 Space Groups and Space Group Types

The Hermann–Mauguin symbol for a space group type begins with a capital letter (*P, A, B, C, F, I* or *R*) which expresses the presence of translation symmetry and the kind of centering. The letter is followed by a listing of the other symmetry elements according to the same rules as for point groups, the base vectors **a**, **b** and **c** defining the coordinate system (Fig. 146).

Symmetry axes can only have the multiplicities 1, 2, 3, 4 or 6 when translation symmetry is present in three dimensions. If, for example, fivefold axes were present in one direction, the unit cell would have to be a pentagonal prism; space cannot be filled, free of voids, with prisms of this kind. Due to the restriction to certain multiplicities, symmetry elements can only be combined in a finite number of ways in the presence of three-dimensional translation symmetry. The 230 combination possibilities are called *space group types*.

The 230 space group types are listed in full in the *International Tables for Crystallography* [55]. Whenever crystal symmetry is to be considered, these tables should be consulted. They include figures that show the relative positions of the symmetry elements and they give details concerning all possible sites in the unit cell.

When an atom is situated on an inversion center, a rotation axis or a reflection plane, it takes a *special position*. If the corresponding symmetry operation is performed, the atom is mapped onto itself. Any other site is a *general position*. A special position has some specific *site symmetry*. Molecules in crystals frequently occupy special positions; this implies that the site symmetry cannot be higher than the symmetry of the free molecule. For example, an octahedral ion such as $SbCl_6^-$ can be placed at a position with site symmetry 4 if its Sb atom and two *trans* Cl atoms are placed on the fourfold axis; a water molecule, however, cannot be placed on a fourfold axis.

In some circumstances the magnitudes of the translation vectors must be taken into account. Let us demonstrate this with the example of the trirutile structure. If we triplicate the unit cell of rutile in the **c** direction, we can occupy the metal atom positions with two kinds of metals in a ratio of 1:2, such as is shown in Fig. 147. This structure type is known for several oxides and fluorides, e.g. $ZnSb_2O_6$. Both the rutile and the trirutile structure belong to the same space group *type* $P4_2/mnm$. Due to the triplicated translation vector in the **c** direction the density of the symmetry elements in trirutile is only $\frac{1}{3}$ of those in rutile; in other words, rutile has a symmetry that is higher by a factor of three. A structure with a specific symmetry *including* the translation symmetry has a specific *space group*; the space group type, however, is independent of the special magnitudes of the translation vectors. Therefore, rutile and trirutile do *not* have the same space group. Although space group and space group type have to be distinguished, the same symbols are used for both. However, this does not cause any problems since the specification of a space group is only used to designate the symmetry of a specific structure or a specific structure type, and this always involves a translation lattice with definite dimensions.

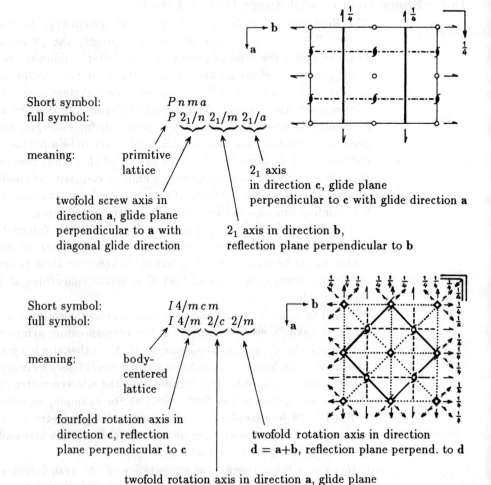

Short symbol: *P n m a*
full symbol: $P\ 2_1/n\ 2_1/m\ 2_1/a$

meaning: primitive
                  lattice

twofold screw axis in
direction **a**, glide plane
perpendicular to **a** with
diagonal glide direction

2_1 axis in direction **b**,
reflection plane perpendicular to **b**

2_1 axis
in direction **c**, glide plane
perpendicular to **c** with glide direction **a**

Short symbol: *I 4/m c m*
full symbol: $I\ 4/m\ 2/c\ 2/m$

meaning: body-
                  centered
                  lattice

fourfold rotation axis in
direction **c**, reflection
plane perpendicular to **c**

twofold rotation axis in direction **a**, glide plane
with glide direction **c** perpendicular to **a**

twofold rotation axis in direction
d = **a**+**b**, reflection plane perpend. to **d**

Short symbol: $P 2_1/c$
full symbol $P 1\ 2_1/c\ 1$

meaning: primitive
                  lattice

no symmetry elements
in directions **a** and **c**

2_1 axis in direction **b**, glide plane
perpendicular to **b** with glide direction **c**

Fig. 146
Examples of
space group
type symbols
and their
meanings

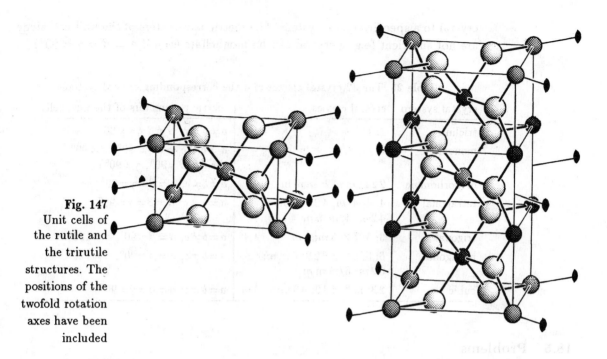

Fig. 147
Unit cells of
the rutile and
the trirutile
structures. The
positions of the
twofold rotation
axes have been
included

18.4 Crystal Classes and Crystal Systems

A well-grown crystal exhibits a macroscopic symmetry which is apparent from its faces; this symmetry is intimately related to the pertinent space group. Due to its finite size, a macroscopic crystal can have no translation symmetry. Its exterior symmetry is that of the point group resulting from the corresponding space group if translation symmetry is removed, screw axes are replaced by rotation axes, and glide planes are replaced by reflection planes. This way the 230 space group types can be correlated with 32 point groups which are called *crystal classes*. Examples for some space group types and the crystal classes to which they belong are:

space group type		crystal class	
full symbol	short symbol	full symbol	short symbol
$P\,1\,2_1/c\,1$	$P\,2_1/c$	$1\,2/m\,1$	$2/m$
$C\,2/m\,2/c\,2_1/m$	$C\,m\,c\,m$	$2/m\,2/m\,2/m$	$m\,m\,m$
$P\,6_3/m\,2/m\,2/c$	$P\,6_3/m\,m\,c$	$6/m\,2/m\,2/m$	$6/m\,m\,m$

in general: the P, A, B, C, F, I or R of the space group symbol is removed, the subscript numbers are removed, and a, b, c, n or d are replaced by m

A special coordinate system defined by the base vectors **a**, **b** and **c** belongs to each space group. Depending on the space group, certain relations hold among the base vectors; they serve to classify seven *crystal systems*. Every crystal class can be assigned to one of these crystal systems, as listed in Table 25. The existence of the corresponding symmetry elements is relevant for assigning a crystal to a specific crystal system. The metric parameters of the unit cell alone are not sufficient (e.g. a crystal can be monoclinic even if $\alpha = \beta = \gamma = 90°$).*

Table 25 The 32 crystal classes and the corresponding crystal systems

crystal system	crystal classes	metric parameters of the unit cell
triclinic	$1; \bar{1}$	$a \neq b \neq c; \ \alpha \neq \beta \neq \gamma \neq 90°$
monoclinic	$2; m; 2/m$	$a \neq b \neq c; \ \alpha = \gamma = 90°, \ \beta \neq 90°$ (or $\alpha = \beta = 90°, \ \gamma \neq 90°$)
orthorhombic	$222; mm2; mmm$	$a \neq b \neq c; \ \alpha = \beta = \gamma = 90°$
tetragonal	$4; \bar{4}; 4/m; 422; 4mm;$ $\bar{4}2m; 4/mmm$	$a = b \neq c; \ \alpha = \beta = \gamma = 90°$
trigonal	$3; \bar{3}; 32; 3m; \bar{3}m$	$a = b \neq c; \ \alpha = \beta = 90°, \ \gamma = 120°$
hexagonal	$6; \bar{6}; 6/m; 622; 6mm;$ $\bar{6}2m; 6/mmm$	$a = b \neq c; \ \alpha = \beta = 90°, \ \gamma = 120°$
cubic	$23; m3; 432; \bar{4}3m; m\bar{3}m$	$a = b = c; \ \alpha = \beta = \gamma = 90°$

18.5   Problems

18.1 Give the Hermann–Mauguin symbols for the following molecules or ions: H_2O, $HCCl_3$, BF_3, XeF_4, $ClSF_5$, SF_6, *cis*-$SbF_4Cl_2^-$, *trans*-N_2F_2, $B(OH)_3$ (planar), $Co(NO_2)_6^{3-}$.

18.2 Plots of the following molecules or ions can be found on pp. 121 and 122. State their Hermann–Mauguin symbols.
Si_4^{6-}, As_4S_4, P_4S_3, Sn_5^{2-}, Sn_9^{4-}, As_4^{6-}, Sb_6^{8-}, As_4^{4-}, P_6^{6-}, P_7^{3-}, P_{11}^{3-}.

18.3 What Hermann–Mauguin symbols correspond to the linked polyhedra shown in Fig. 87 (p. 157)?

18.4 What symmetry elements are present in the HgO chain shown on p. 222?

18.5 Find out which symmetry elements are present in the structures of the following compounds. Derive the Hermann–Mauguin symbol of the corresponding space group (it may be helpful to use the *International Tables for Crystallography*, Vol. A).
Tungsten bronzes M_xWO_3 (Fig. 92, p. 163); NiAs (Fig. 119, p. 192); $BaTiO_3$ (Fig. 135, p. 214).

18.6 State the crystal classes and crystal systems to which the following space groups belong.
(a) $P2_1/b2_1/c2_1/a$; (b) $I4_1/amd$; (c) $R\bar{3}2/m$; (d) $C2/c$; (e) $P6_3/m$; (f) $P6_322$; (g) $P2_12_12_1$.

---

*The classification of the crystal systems is not used in the same manner internationally. In American and Russian literature usually only six crystal systems are distinguished, in that all crystal classes termed hexagonal and trigonal in Table 25 are assigned to the hexagonal system. In the French literature and, sporadically, in German literature rhombohedral is considered to be a crystal system of its own (crystal classes 3, $\bar{3}$, 32, 3m and $\bar{3}m$ if there is rhombohedral centering). The classification given in Table 25 is that of the *International Tables for Crystallography*.

19 Symmetry as the Organizing Principle for Crystal Structures

19.1 Crystallographic Group–Subgroup Relations

A space group consists of a set of symmetry elements that always fulfils the conditions according to which a group is defined in mathematics (cf. e.g. [31, 53,54]). Group theory offers a mathematically clear-cut and very powerful tool for ordering the multitude of crystal structures according to their space groups. To this end we introduce some concepts without discussing group theory itself in detail.

A space group G_1 consists of a set of symmetry elements. If another space group G_2 consists of a subset of these symmetry elements, it is a *subgroup* of G_1; at the same time G_1 is a *supergroup* of G_2. The symmetry elements present in the space group G_1 multiply an atom which is placed in a general position by a factor n_1. A corresponding atom in the subgroup G_2 is multiplied by analogy by a factor n_2. Since G_2 possesses less symmetry elements than G_1, $n_1 > n_2$ holds. The fraction n_1/n_2 is the *index* of the symmetry reduction from G_1 to G_2. It always is an integer and serves to order space groups hierarchically. For example, when passing from rutile to trirutile (cf. the end of section 18.3), there is a symmetry reduction of index 3.

G_2 is a *maximal subgroup* of G_1 if there exists no space group that can act as intermediate group between G_1 and G_2. G_1 then is a *minimal supergroup* of G_2. The index of symmetry reduction from a group to a maximal subgroup always is a prime number or a prime number power. According to the theorem of C. HERMANN a maximal subgroup either is *translationengleich* or *klassengleich*.*

Translationengleiche subgroups have an unaltered translation lattice, i.e. the translation vectors and therefore the size of the primitive unit cells of group and subgroup coincide. The symmetry reduction in this case is accomplished by the loss of other symmetry elements or by the reduction of the multiplicity of symmetry axes. This implies a transition to a different crystal class. The example on the right in Fig. 148 shows how a fourfold rotation axis is converted to a twofold rotation axis when four symmetry-equivalent atoms are replaced by two pairs of different atoms; the translation vectors are not affected.

A group and a *klassengleiche* subgroup belong to the same crystal class. The symmetry reduction takes place by the loss of translation symmetry, i.e. by enlargement of the unit cell or by the loss of centering of the unit cell. With

---

*Terms taken from German: *translationengleich* = with the same translations; *klassengleich* = of the same class.

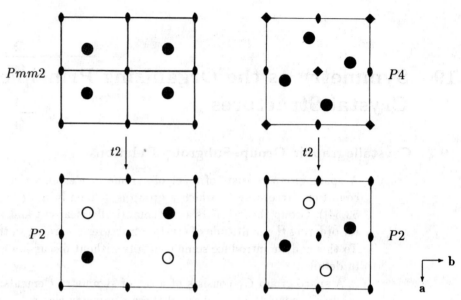

Fig. 148

Examples for *translationengleiche* group–subgroup relations: left, loss of reflection planes; right, reduction of the multiplicity of a rotation axis from 4 to 2. The circles with the same color, ○ and ●, respectively, designate symmetry-equivalent positions

the enlargement of the unit cell the number of other symmetry elements is also reduced, a reflection plane can be converted to a glide plane, and a rotation axis can be converted to a screw axis. Fig. 149 shows two examples.

A special case of *klassengleiche* subgroups are the *isomorphic* subgroups. Group and subgroup belong to the same space group type, and they thus have the same Hermann–Mauguin symbol (for this reason they have also been termed isosymbolic subgroups). The subgroup has an enlarged unit cell. Rutile and trirutile offer an example (Fig. 147).

A suitable way to represent group–subgroup relations is by means of a genealogical tree which shows the relations from groups to their maximal subgroups by arrows pointing downwards. At every arrow the kind of the relation and the index of the symmetry reduction are labeled, e.g.:

$t2$ = *translationengleiche* subgroup of index 2

$k2$ = *klassengleiche* subgroup of index 2

$i3$ = isomorphic subgroup of index 3

In addition, if applicable, it is stated how the new unit cell emerges from the old one (base vectors of the subgroup given as vectorial sums of the base vectors of the preceding group). See Fig. 149.

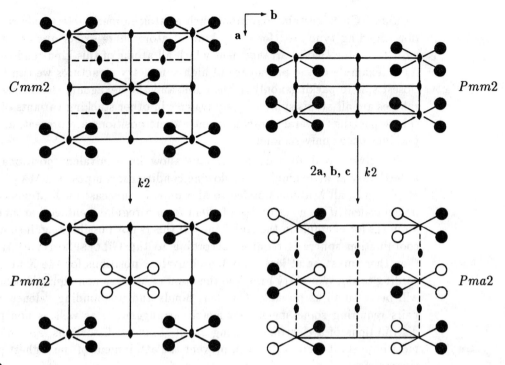

Fig. 149
Examples for *klassengleiche* group–subgroup relations: left, loss of centering including the loss of glide planes and twofold axes; right, enlargement of the unit cell including the conversion of a reflection plane to a glide plane. Circles with the same color, ○ and ●, respectively, refer to symmetry equivalent positions

19.2 The Symmetry Principle in Crystal Chemistry

In crystalline solids a tendency to form arrangements of high symmetry is observable. The *symmetry principle*, put forward in this form by F. LAVES [56,57], has been stated in a more exact manner by H. BÄRNIGHAUSEN [132]:

1. In the solid state the arrangement of atoms shows a pronounced tendency towards the highest possible symmetry.

2. Several counteracting factors may prevent the attainment of the highest possible symmetry, but in most cases the deviations from the ideal arrangement are only small, and frequently the observed symmetry reduction corresponds to the smallest possible step.

3. During a solid state reaction which results in one or more products of lower symmetry, very often the higher symmetry of the starting material is indirectly preserved by the orientation of domains formed within the crystalline matrix.

The background of the symmetry principle can be expressed by the formulation of G. O. BRUNNER [133]: *atoms of the same kind tend to be in equivalent positions.* Given certain conditions such as stoichiometry, electronic configuration, bonding type etc., for every kind of atom there exists *one* energetically most favorable spatial arrangement which all atoms of this kind tend to adopt. As an example of the prevalence of high-symmetry structures we can take the closest sphere packings: only in the cubic and the hexagonal closest-packing of spheres are all atoms symmetry equivalent; in other stacking variants of closest sphere packings several non-equivalent atomic positions are present, and these packings occur only seldomly.

The given conditions do not always allow for equivalent positions for all atoms. Take as an example the following conditions: composition MX_5, covalent M–X bonds, all X atoms bonded to M atoms. In this case all X atoms can only be equivalent if each set of five of them form a regular pentagon around an M atom (as for example in the XeF_5^- ion). If this is not possible (e.g. if some other coordination sphere is required according to the GILLESPIE–NYHOLM rules), then there must be at least two non-equivalent positions for the X atoms. According to the symmetry principle the number of these non-equivalent positions will be as small as possible. Covalent bonds and nonbonding valence electron pairs requiring some specific geometric arrangement, as well as nonspherical distributions of *d* electrons causing JAHN–TELLER distortions, are among the most important factors that can prevent the attainment of the highest possible symmetry.

19.3   Structural Relationships by Group–Subgroup Relations

As was shown in various previous chapters, many structures of solids can be regarded as derivatives of simple, high-symmetry structure types. Let us recall some examples:

Body-centered cubic sphere packing $\Rightarrow$ CsCl type $\Rightarrow$ superstructures of the CsCl type (section 14.3)

Diamond $\Rightarrow$ sphalerite $\Rightarrow$ chalcopyrite (sections 11.2 and 11.4)

Closest-packings $\Rightarrow$ closest-packings with occupied octahedral interstices (e.g. CdI_2 type) (section 16.3)

In all cases we start from a simple structure which has high symmetry. Every arrow ($\Rightarrow$) in the preceding examples marks a reduction of symmetry, i.e. a group–subgroup relation. Since these are well-defined mathematically, they are an ideal tool for revealing structural relationships in a systematic way. Changes that may be the reason for symmetry reductions include:

- Atoms of an element in symmetry-equivalent positions are substituted by several kinds of atoms. Example: CC (diamond) $\rightarrow$ ZnS (sphalerite).

- Atoms are replaced by voids or voids are occupied by atoms. Example: hexagonal closest-packing $\rightarrow$ CdI_2 type. If the voids are considered to be "zero atoms", this can be considered as a "substitution" of voids by atoms.

- Atoms of an element are substituted by atoms of another element that requires an altered coordination geometry. For example: $KMgF_3$ (perovskite type) $\rightarrow$ $CsGeCl_3$ (lone electron pair at the Ge atom, Ge atom shifted from the octahedron center towards an octahedron face so that the three covalent bonds of an $GeCl_3^-$ ion are formed); $CdBr_2$ (CdI_2 type) $\rightarrow$ $CuBr_2$ (CdI_2 type distorted by the JAHN–TELLER effect).

- Emergence of new interactions. For example: iodine (high pressure, metallic, sphere packing) $\rightarrow$ I_2 molecules (normal pressure).

Structural relations can be presented in a clear and concise manner using family trees of group–subgroup relations as put forward by BÄRNIGHAUSEN [132]. They can be set up with the aid of the *International Tables for Crystallography* [55], in which the maximal subgroups of every space group are listed. The top of the family tree ("BÄRNIGHAUSEN tree") corresponds to the structure of highest symmetry, the *aristo type*, from which all other structure types of a structure family are derived. Every space group listed in the family tree represents one structure type. Since the space group symbol itself states only symmetry, and gives no information about the atomic positions, additional information concerning these is necessary for every member of the family tree (atomic coordinates, site symmetries, number of symmetry-equivalent positions). In simple cases these data can be included in the family tree; in more complicated cases an additional table is convenient.

Inspection of the atomic positions reveals how the symmetry is being reduced stepwise. In the aristo type usually all atoms are situated in special positions, i.e. they have positions on certain symmetry elements, fixed values of the coordinates, and specific site symmetries. From group to group the following changes occur with the atomic positions: 1. Individual values of the coordinates x, y, z become independent, i.e. the atoms can shift away from the fixed values of a special position. 2. The site symmetry decreases. 3. Symmetry-equivalent positions split into several positions independent from one another. The WYCKOFF symbol is commonly used to designate an independent position, for example $4c$; the 4 specifies the existence of four symmetry-equivalent sites for the corresponding position within the unit cell, and c is merely an alphabetical label (a, b, c, ...) according to the listing of the positions in the *International Tables for Crystallography*. The following example (family tree 1) shows how specifications can be made for the site occupations; in the remaining examples (family trees 2 to 4) no detailed data are given, but instead the positions of interest are marked by boxes with different gray tones.

Diamond–Sphalerite

The group–subgroup relation of the symmetry reduction from diamond to sphalerite is shown in family tree 1. Some comments concerning the terminology have been included. The corresponding structures are shown on pp. 107 and 108. In both structures the atoms have identical coordinates and site symmetries. The

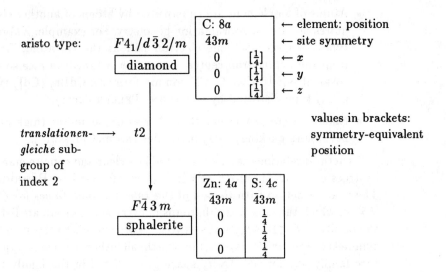

Family tree 1
Group–subgroup relation diamond–sphalerite. The numbers in the boxes are atomic coordinates. Atomic positions are labeled by WYCKOFF symbols, e.g. 8a

unit cell of diamond contains eight C atoms in symmetry-equivalent positions (position 8a); with the symmetry reduction the atomic positions split to two independent positions (4a and 4c) which are occupied in sphalerite by four zinc atoms and four sulfur atoms. The space groups are *translationengleich*: the dimensions of the unit cells correspond to each other. The symmetry reduction involves a loss of the inversion centers which in diamond are present in the centers of the C–C bonds.

Occupation of Octahedral Interstices in Hexagonal Closest-packing

According to the discussion in section 16.3, many structures can be derived from the hexagonal closest-packing of spheres by occupying a fraction of the octahedral interstices with other atoms. If the X atoms of a compound MX_n form the sphere packing, then the fraction $1/n$ of the octahedral interstices must be occupied. The unit cell of the hexagonal closest-packing contains two octahedral interstices, so that only the fractions $\frac{1}{1}$ or $\frac{1}{2}$ of the octahedral interstices can be occupied without enlargement of the unit cell; occupation of any other fraction requires an enlargement of the unit cell. In other words, only the compositions MX and MX_2 allow for structures without cell enlargement. Cell enlargement is equivalent to loss of translation symmetry, therefore *klassengleiche* group–subgroup relations must occur.

The aristo type can be considered to be either the sphere packing itself, or the NiAs type which corresponds to the sphere packing in which all octahedral interstices are occupied by Ni atoms. In the aristo type these interstices are

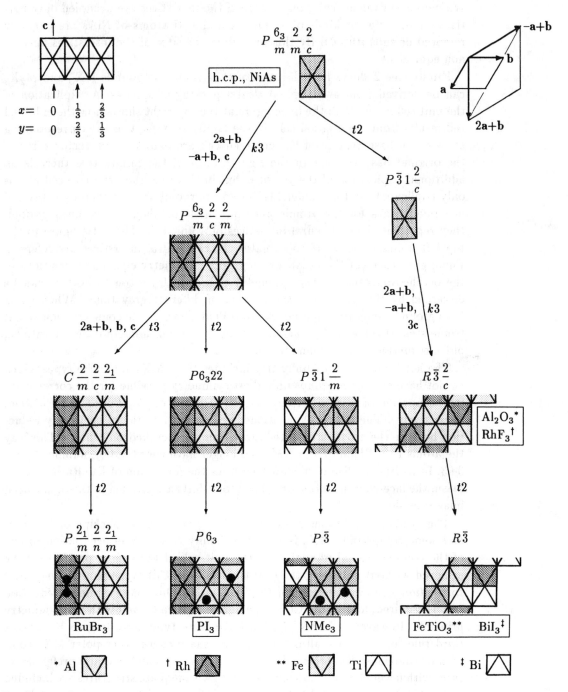

Family tree 2

Group–subgroup relations from hexagonal closest-packing to some MX_3 and M_2X_3 structures

symmetry-equivalent; subgroups result if the interstices are occupied only partially or by different kinds of atoms (or if the Ni atoms of NiAs are partially removed or substituted). By this procedure the sites of the interstices become non-equivalent.

Family tree 2 shows how the structures of some compounds MX_3 and M_2X_3 can be derived from a hexagonal closest-packing of spheres. A triplication of the unit cell is required; the little figure at the top right shows how the enlarged cell results from the original cell. In the family tree, next to the corresponding arrows, the base vectors of the enlarged cell are expressed as vector sums of the original base vectors. In the right branch of the family tree there is an additional triplication of the vector **c**, but in this case the primitive cell also is only triplicated (and not ninefold) due to the rhombohedral centering. Instead of numeric data for the atomic positions, little octahedra have been plotted; they represent the octahedral interstices of a unit cell. The little figure to the top left indicates to what coordinates the octahedral interstices are referred. Equal gray tones of the octahedra represent symmetry-equivalent positions of the octahedron centers. The symmetry reduction from top to bottom can be discerned by the increase of the number of different gray tones. When going from group to subgroup, symmetry-equivalent octahedra become non-equivalent (some steps involve not an increase of the number of non-equivalent positions, but a reduction of site symmetries).

The left branch of the family tree includes three MX_3 structure types which do not have the space groups of highest symmetry possible for the corresponding occupation of octahedral interstices ($P6_3/m\,2/c\,2/m$, $P6_322$ and $P\bar{3}12/m$, respectively). The examples mentioned require an additional symmetry reduction because the atoms are shifted away from the octahedron centers. This way the atoms P in PI_3 and N in NMe_3 no longer have coordination number 6, but 3+3. In $RuBr_3$ the Ru atom shift facilitates the formation of Ru–Ru bonds between the face-sharing octahedra. The atom shifts are marked in the octahedron images by dots.

The symmetry relations between the hexagonal closest-packing of X atoms and some compounds MX_2 is shown in family tree 3. The occupation of one of the two octahedral interstices of the unit cell of the sphere packing can be achieved without enlarging the cell: the resulting CdI_2 type thus belongs to a *translationengleiche* subgroup of $P6_3/m\,2/m\,2/c$. Only one symmetry reduction step is required, and so the CdI_2 type agrees especially well with the symmetry principle. However, it is not the ideal structure type according to PAULING's third rule (p. 45), since all octahedra only share edges. Very polar MX_2 compounds (dioxides, difluorides, dichlorides of alkaline earth metals) prefer structures with fewer shared octahedron edges. Two adequate structures are included in the family tree, the α-PbO_2 and the $CaCl_2$ type. They are placed at lower levels in the symmetry hierarchy than the CdI_2 type since the X atoms occupy two kinds of non-equivalent positions. The $CaCl_2$ type can also be derived from the rutile type (cf. p. 194) which is the aristo type of another structure family

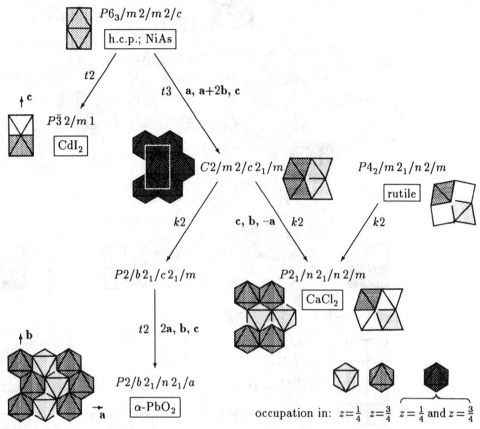

Family tree 3

Group–subgroup relations from hexagonal closest-packing of spheres and from the rutile type to some MX_2 structures. Cf. Fig. 98, p. 168 and Fig. 120, p. 194. Gray boxes represent occupied octahedral interstices; they are symmetry-equivalent irrespective of the gray tone

(not shown). Unlike in family tree 2, different gray tones of the boxes represent not different symmetry-independent octahedron sites, but different heights in the direction of view.

Occupation of Tetrahedral Interstices in Cubic Closest-packing

Relations concerning some structures that can be derived from the CaF_2 type (cubic closest-packing of spheres with occupation of all tetrahedral interstices) are shown in family tree 4. Starting from CaF_2, atoms are removed successively from the tetrahedral interstices. Compare the corresponding structures shown in Fig. 128 (p. 203). Like family tree 3, family tree 4 also includes some steps of symmetry reduction which do not alter the distribution of equivalent positions, but only involve a reduction of the site symmetries. The stated space group symbols do not correspond in all cases to the standard settings of the *Interna-*

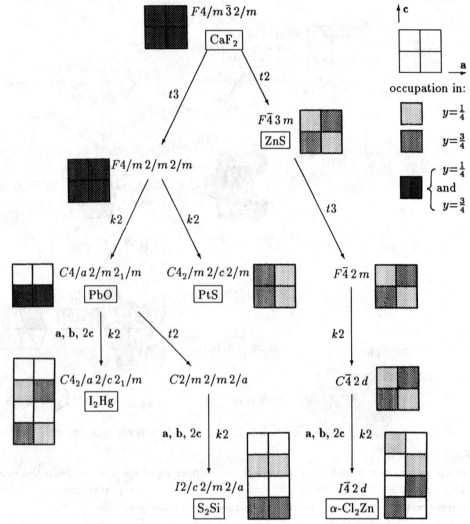

Family tree 4

Group–subgroup relations for some structures that can be derived from the CaF_2 type by succesive removal of atoms from tetrahedral interstices. The boxes represent tetrahedral interstices, and the gray tone marks their occupation as depicted at the top right (all occupied tetrahedra are symmetry-equivalent). To avoid changes in the cell orientations, non-standard settings have been used for some tetragonal space groups (standard setting in parentheses, having base vectors $\frac{1}{2}(\mathbf{a+b})$, $\frac{1}{2}(\mathbf{a-b})$, $\mathbf{c}$): $F4/m\,2/m\,2/m$ $(I4/m\,2/m\,2/m)$; $C4/a\,2/m\,2_1/m$ $(P4/n\,2_1/m\,2/m)$; $C4_2/m\,2/c\,2/m$ $(P4_2/m\,2/m\,2/c)$; $C4_2/a\,2/c\,2_1/m$ $(P4_2/n\,2_1/m\,2/c)$; $F\bar{4}\,2\,m$ $(I\bar{4}\,m\,2)$; $C\bar{4}\,2\,d$ $(P\bar{4}\,n\,2)$

tional Tables; this way unit cell transformations have been avoided which could cause some confusion. Also, the choice of the unit cell origins of Fig. 128 does not correspond in all cases to standard conventions.[*]

---

[*]Except for CaF_2 the conventional cell origin is taken to be in one of the tetrahedrally coordinated atoms or in an unoccupied inversion center.

19.4 Twinned Crystals

Twinned crystals consist of crystals intergrown in a symmetrical manner. They may consist of individuals that can be depicted macroscopically as in the case of the "dovetail twins" of gypsum, where the two components are mirror-inverted (Fig. 150). There may also be numerous alternating components which sometimes cause a streaky appearance of the crystals (polysynthetic twin). One of the twin components is converted to the other by some symmetry operation (twinning operation), for example by a reflection in the case of the dovetail twins. Another example is the "Dauphiné twins" of quartz which are interconverted by a twofold rotation axis (Fig. 150). Threefold or fourfold axes can also occur as symmetry elements between the components; the crystals then are triplets or quadruplets.

Fig. 150
(a) Dovetail twin (gypsum).
(b) Polysynthetic twin (felspar).
(c) Dauphiné twin (quartz)

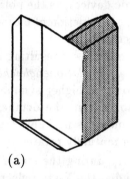 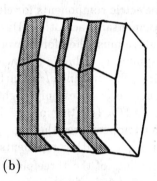 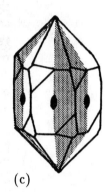

(a)   (b)   (c)

At the interface between the twin components there is a symmetry element which is not a symmetry element of the structure itself, but which must be compatible with the given structural facts. For example: if a stacking fault occurs in a cubic closest-packing of spheres,

$$\ldots A_B C^A B_C^| B^A C B^A \ldots$$

twins will result which have a reflection plane perpendicular to the stacking direction. The cubic closest-packing itself has no reflection planes with this orientation, but as a twinning operation the reflection is compatible with the packing.

During a phase transition of one structure to another structure having lower symmetry, twins, triplets or quadruplets will always be formed if the transition includes a *translationengleiche* group–subgroup relation. The index of this relation determines the kind of polytuple. An example is the Dauphiné twins of quartz which are formed when quartz is transformed from its high-temperature form (β or high quartz, stable above of 573 °C) to the low-temperature form (α or low quartz). In the following images only the silicon atoms around a screw axis are shown (O atoms are situated on the connecting lines):

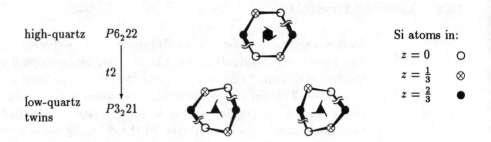

high-quartz $P6_222$

$t2$

low-quartz
twins $P3_221$

Si atoms in:

$z = 0$ ○
$z = \frac{1}{3}$ ⊗
$z = \frac{2}{3}$ ●

The space group of low quartz is a *translationengleiche* subgroup of index 2 of the space group of high quartz, and twins are the result. During the symmetry reduction certain twofold axes are lost (the last 2 in the space group symbol $P6_222$). These axes, present only in the higher symmetry space group, produce the twinning operation. Quartz crystals twinned in this manner are unsuitable as piezoelectric components for electronic devices, as the polar directions of the twins compensate each other. During the production of piezoelectric quartz the temperature must therefore never exceed 573 °C.

The symmetry reduction shown in family tree 2 resulting in the space group of RuBr$_3$ ($P2_1/m\,2/n\,2_1/m$) includes a *translationengleiche* step of index 3. The structure deviates only slightly from the higher symmetrical space group $P6_3/m\,2/c\,2/m$ (shifting of the Ru atoms from the octahedron centers) and at higher temperatures the higher symmetry is probably true. RuBr$_3$ crystals occur as triplets with components being mutually rotated by 120°, corresponding to the loss of the threefold symmetry during the step $P6_3/m\,2/c\,2/m \overset{t3}{\to} C2/m\,2/c\,2_1/m$. In X-ray diffraction studies the X-ray reflections of the triplets appear superimposed, so that they give the impression of hexagonal symmetry and the resulting Ru atom positions seem to be exactly in the octahedron centers.

19.5 Problems

19.1 Determine whether the following group–subgroup relations are *translationengleich*, *klassengleich* or isomorphic. If the unit cell of the subgroup is enlarged, this is stated on top of the arrow.

(a) $Cmcm \rightarrow Pmcm$; (b) $P2_1/c \rightarrow P\bar{1}$; (c) $Pbcm \overset{2a,b,c}{\longrightarrow} Pbca$; (d) $C12/m1 \overset{a,3b,c}{\longrightarrow} C12/m1$; (e) $P6_3/mcm \rightarrow P6_322$; (f) $P2_1/m2_1/m2/n \rightarrow Pmm2$; (g) $P2_1/m2_1/m2/n \rightarrow P12_1/m1$.

19.2 The phase transition from α-tin to β-tin involves a change of space group $F4_1/d\bar{3}2/m \overset{t3}{\to} I4_1/a2/m2/d$. Would you expect β-tin to form twins or multiplets?

19.3 The phase transition of NaNO$_2$ at 164 °C from the paraelectric to the ferroelectric form involves a change of space group from $I2/m2/m2/m$ to $Imm2$. Will the ferroelectric phase be twinned?

References

General Literature, Text Books

[1] D.F. Shriver, P.W. Atkins and C.H. Langford, *Inorganic Chemistry*. Oxford University Press, 1990.

[2] J.D. Lee, *Concise Inorganic Chemistry*, 4th ed. Chapman & Hall, London 1991.

[3] J.E. Huheey, *Inorganic Chemistry*, 3rd ed. Harper, Cambridge 1983.

[4] N.N. Greenwood and A. Earnshaw, *Chemistry of the Elements*. Pergamon, Oxford 1984.

[5] F.A. Cotton and G.Wilkinson, *Advanced Inorganic Chemistry*, 5th ed. J. Wiley, New York 1988.

[6] C. Elschenbroich and A. Salzer, *Organometallics*. VCH Publishers, Weinheim 1989.

[7] A.R. West, *Solid State Chemistry*, Clarendon, Oxford 1992.

[8] L. Smart and E. Moore, *Solid State Chemistry, an Introduction*, Chapman & Hall, London 1992.

[9] A.K. Cheetham, *Solid State Chemistry and its Applications*, J. Wiley, London 1984.

[10] A.F. Wells, *Structural Inorganic Chemistry*, 5th ed. Clarendon, Oxford 1984.

[11] H. Krebs, *Fundamentals of Inorganic Crystal Chemistry*. McGraw-Hill, London 1968.

[12] R.C. Evans, *Crystal Chemistry*. Cambridge Universtity Press 1966.

[13] H.W. Jaffe, *Introduction to Crystal Chemistry*. Cambridge University Press, 1989.

[14] G.M. Clark, *The Structures of Non-molecular Solids*. Applied Science Publishers, London 1972.

[15] D.M. Adams, *Inorganic Solids*. J. Wiley, London 1974.

[16] N.N. Greenwood, *Ionic Crystals, Lattice Defects and Nonstoichiometry*. Butterworths, London 1968.

[17] B.K. Vainshtein, V.M. Fridkin and V.L. Indenbom, *Modern Crystallography II: Structure of Crystals*. Springer, Heidelberg 1982.

[18] K. Schubert, *Kristallstrukturen zweikomponentiger Phasen*. Springer, Berlin 1964.

[19] B.G. Hyde and S. Andersson, *Inorganic Crystal Structures*. J. Wiley, New York 1989.

[20] H.D. Megaw, *Crystal Structures, a Working Approach*. W.B. Saunders, Philadelphia 1973.

[21] R.J. Gillespie, *Molecular Geometry*. Van Nostrand Reinhold, London 1972.

[22] R.J. Gillespie and I. Hargittai, *VSEPR Model of Molecular Geometry*. Allyn and Bacon, Boston 1991.

[23] L. Pauling, *The Nature of the Chemical Bond*, 3rd ed. Cornell University Press 1960.

[24] D.L. Kepert, *Inorganic Stereochemistry*. Springer, Berlin 1982.

[25] A. Weiss and H. Witte, *Kristallstruktur und chemische Bindung*. Verlag Chemie, Weinheim 1983.

[26] S.F.A. Kettle, *Symmetry and Structure*. J. Wiley, New York 1985.

[27] C. Hammond, *Introduction to Crystallography*. Oxford University Press, 1990.

[28] H.L. Monaco, D.Viterbo, F.Sordari, G.Grilli, G. Zanotti and M. Catti, *Fundamentals of Crystallography*. Oxford University Press, 1992.

[29] W. Kleber, H.H Bautsch and J. Bohm, *Einführung in die Kristallographie*. 17th ed. Verlag Technik, Berlin 1990.

[30] W. Borchardt-Ott, *Kristallographie, eine Einführung für Naturwissenschaftler*, 3rd. ed. Springer, Berlin 1990.

[31] H. Burzlaff and H. Zimmermann, *Kristallographie*. G. Thieme, Stuttgart 1977.

Monographs to Single Topics

(Sequence according to the sequence of chapters 8–19)

[32] C.J. Ballhausen, *Ligand Field Theory*. McGraw-Hill, New York 1962.

[33] B.N. Figgis, *Introduction to Ligand Fields*. Wiley, London–New York 1966.

[34] H.L. Schläfer and C. Gliemann, *Basic Principles of Ligand Field Theory*. Wiley, New York 1969.

[35] Y. Saito, *Inorganic Molecular Dissymmetry*. Springer, Berlin 1979.

[36] A.I. Kitaigorodsky, *Molecular Crystals and Molecules*. Academic Press, New York, 1973.

[37] R. Hoffmann, *Solids and Surfaces. A Chemist's View of Bonding in Extended Structures*. VCH Publishers, Weinheim, 1988.

[38] P.A. Cox, *The Electronic Structure and Chemistry of Solids*. Clarendon, Oxford 1987.

[39] J.A. Duffy, *Bonding, Energy Levels and Bands in Inorganic Chemistry*. Longman Scientific & Technical, 1990.

[40] S.L. Altmann, *Band Theory of Solids. An Introduction from the Point of View of Symmetry*. Clarendon, Oxford 1991.

[41] H. Jones, *The Theory of Brillouin Zones and Electronic States in Crystals*. North Holland, Amsterdam 1962.

[42] H. Zabel and S.A. Solin, *Graphite Intercalation Compounds I; Structure and Dynamics*. Springer, Berlin 1990.

[43] W. Hume-Rothery, R. E. Sallmann and C. W. Haworth, *The Structures of Metals and Alloys*, 5th ed. Institute of Metals, London 1969.

[44] W.B. Pearson, *The Crystal Chemistry and Physics of Metals and Alloys*. J. Wiley, New York 1972.

[45] O. Muller and R. Roy, *The Major Ternary Structural Families*. Springer, Berlin 1974.

[46] D.J.M. Bevan and P. Hagenmuller, *Nonstoichiometric Compounds: Tungsten Bronzes, Vanadium Bronzes and Related Compounds*. Pergamon, Oxford 1975.

[47] E. Parthé, *Crystallochimie des Structures Tétraédriques*. Gordon & Breach, Paris 1972.

[48] F. Liebau, *Structural Chemistry of Silicates* . Springer, Berlin 1985.

[49] D.W. Breck, *Zeolite Molecular Sieves*. J. Wiley, New York, 1974.

[50] F.E. Mabbs, *Magnetism and Transition Metal Complexes*. Chapman & Hill, London 1973.

[51] R.L. Carlin, *Magnetochemistry*. Springer, Berlin 1986.

[52] B. Vainshtein, *Modern Crystallography I: Symmetry of Crystals*. Springer, Berlin 1981.

[53] W. Ledermann, *Introduction to Group Theory*. Longman, London 1976.

[54] F.A. Cotton, *Chemical Applications of Group Theory*, 3rd ed. J. Wiley, New York 1990.

[55] *International Tables for Crystallography*, Vol. A, 2nd ed. (T. Hahn, editor). Kluwer, Dordrecht 1987.

[56] F. Laves, Crystal Structure and Atomic Size. In: *Theory of Alloy Phases*, p. 124. American Society for Metals, Cleveland 1956.

[57] F. Laves, *Phase Stability in Metals and Alloys*. McGraw-Hill, New York 1967.

Collections of Molecular and Crystal Structure Data

[58] R.W.G. Wyckoff, *Crystal Structures*, Vols 1–6. J. Wiley, New York 1962–1971.

[59] J. Donohue, *The Structures of the Elements*. J. Wiley, New York 1974.

[60] *Strukturbericht* 1–7. Akad. Verlagsges., Leipzig 1931–1943. *Structure Reports* 8ff. Kluwer, Dordrecht 1956ff. Collection of crystal structure data of one year, published annually.

[61] *Molecular Structures and Dimensions* (O. Kennard and others, editors). D. Reidel, Dordrecht 1970ff.

[62] Landolt-Börnstein, *Zahlenwerte und Funktionen aus Naturwiss. und Technik*, Neue Serie (K.H. Hellwege, editor), III, Vol. 7. Springer, Berlin 1973–1978.

[63] P. Villars and L.D. Calvert, *Pearson's Handbook of Crystallographic Data for Intermetallic Phases*, 2nd ed., Vols 1–4. ASM International, Materials Park, Ohio 1991.

[64] J.L.C. Daams, P. Villars and J.H.N. van Vucht, *Atlas of Crystal Structure Types for Intermetallic Phases*, Vols 1–4. ASM International, Materials Park, Ohio 1991.

[65] Molecular Gas Phase Documentation (MOGADOC). Fachinformationszentrum Karlsruhe, Germany. Continuously updated data base about molecular structures in the gas phase.

[66] Inorganic Crystal Structure Data Base (ICSD). Fachinformationszentrum Karlsruhe, Germany. Continuously updated data base about crystal structures of inorganic compounds.

[67] Cambridge Structural Data Base (CSD). Cambridge Crystallographic Data Centre, University Chemical Laboratory, Cambridge, England. Continuously updated data base about crystal structures of organic and organometallic compounds.

[68] Metals Crystallographic Data File (CRYSTMET). National Research Council of Canada, Ottawa. Continuously updated data base about crystal structures of metals and intermetallic compounds including metal sufides, phosphides etc.

Selected Papers in Scientific Journals

[69] A. Simon, Intermetallic Compounds and the Application of Atomic Radii for their Description. *Angew. Chem. Intern. Ed.* **22** (1983) 95.

[70] G.O. Brunner and D. Schwarzenbach, Zur Abgrenzung der Koordinationssphäre und Ermittlung der Koordinationszahl in Kristallstrukturen. *Z. Kristallogr.* **133** (1971) 127.

[71] F.C. Frank and J.S. Kasper, Complex Alloy Structures Regarded as Sphere Packings. I: Definitions and Basic Principles, *Acta Crystallogr.* **11** (1958) 184. II: Analysis and Classification of Representative Structures, *Acta Crystallogr.* **12** (1959) 483.

[72] R. Hoppe, The Coordination Number — an Inorganic Chameleon. *Angew. Chem. Intern. Ed.* **9** (1970) 25.

[73] F.L. Carter, Quantifying the Concept of Coordination Number. *Acta Crystallogr.* **B 34** (1978) 2962.

[74] R. Hoppe, Effective Coordination Numbers and Mean Fictive Ionic Radii. *Z. Kristallogr.* **150** (1979) 23.

[75] J. Lima-de-Faria, E. Hellner, F. Liebau, E. Makovicky and E. Parthé, Nomenclature of Inorganic Structure Types. *Acta Crystallogr.* **A 46** (1990) 1.

[76] E. Parthé and L.M. Gelato, The Standardization of Inorganic Crystal Structure Data. *Acta Crystallogr.* **A 40** (1984) 169.

[77] S.W. Bailey, V.A. Frank-Kamentski, S. Goldsztaub, H. Schulz, H.F.W. Taylor, M. Fleischer and A.J.C. Wilson, Report of the International Mineralogical Association – International Union of Crystallography Joint Committee on Nomenclature. *Acta Crystallogr.* **A 33** (1977) 681.

[78] S. Andersson, A Description of Complex Inorganic Crystal Structures. *Angew. Chem. Intern. Ed.* **22** (1983) 69.

[79] A. Bondi, Van der Waals Volumes and Radii. *J. Phys. Chem.* **68** (1964) 441.

[80] S.C. Nyborg and C.H. Faerman, A Review of van der Waals Atomic Radii for Molecular Crystals. I: N, O, F, S, Se, Cl, Br and I Bonded to Carbon. *Acta Crystallogr.* **B 41** (1985) 274. II: Hydrogen Bonded to Carbon. S.C. Nyborg, C.H. Faerman and L. Prasad, *Acta Crystallogr.* **B 43** (1987) 106.

[81] J. Krug, H. Witte and E. Wölfel, Röntgenographische Bestimmung der Elektronenverteilung in Kristallen. *Z. Phys. Chem. N.F.* **4** (1955) 36.

[82] R.D. Shannon, Revised Effective Ionic Radii and Systematic Studies of Interatomic Distances in Halides and Chalcogenides. *Acta Crystallogr.* **A 32** (1976) 751.

[83] R.D. Shannon, Bond Distances in Sulfides and a Preliminary Table of Sulfide Crystal Radii. In: *Structure and Bonding in Crystals*, Vol. II (M. O'Keeffe and A. Navrotsky, eds.). Academic Press, New York 1981.

[84] W.H. Baur, Bond Length Variation and Distorted Coordination Polyhedra in Inorganic Crystals. *Trans. Amer. Crystallogr. Assoc.* **6** (1970) 129. — Interatomic Distance Predictions for Computer Simulation of Crystal Structures, in: *Structure and Bonding in Crystals*, Vol. II, p. 31 (M. O'Keeffe and M. Navrotsky, eds.). Academic Press, New York 1981.

[85] T.C. Waddington, Lattice Energies and their Significance in Inorganic Chemistry. *Advan. Inorg. Chem. Radiochem.* **1** (1959) 157.

[86] M.F.C. Ladd and W.H. Lee, Lattice Energies and Related Topics, *Progr. Solid State Chem.* (1964) 37; **2** (1965) 378; **3** (1967) 265.

[87] R. Hoppe, Madelung Constants as a new Guide to the Structural Chemistry of Solids. *Adv. Fluor. Chem.* **6** (1970) 387.

[88] R.J. Gillespie, The VSEPR Model Revisited. *J. Chem. Soc. Rev.* **21** (1992) 59.

[89] J.R. Edmundson, On the Distribution of Point Charges on the Surface of a Sphere. *Acta Crystallogr.* **A 48** (1991) 60.

[90] J. Brickmann, M. Klöffler and H.U. Raab, Atomorbitale. *Chemie in uns. Zeit* **12** (1978) 23.

[91] H. Selig and L.B. Ebert, Graphite Intercalation Compounds. *Advan. Inorg. Chem. Radiochem.* **23** (1980) 281.

[92] H.W. Kroto, Buckminsterfullerene, a Celestial Sphere that fell to Earth. *Angew. Chem. Intern. Ed.* **31** (1992) 111.

[93] J.M. Burdett, Perspectives in Structural Chemistry. *Chem. Rev.* **88** (1988) 3.

[94] J.W. Lauher, The Bonding Capabilites of Transition Metals Clusters. *J. Amer. Chem. Soc.* **100** (1978) 5305.

[95] K. Wade, Structural and Bonding Patterns in Cluster Chemistry. *Advan. Inorg. Chem. Radiochem.* **18** (1976) 1.

[96] D.M.P. Mingos, T. Slee and L. Zhenyang, Bonding Models for Ligated and Bare Clusters. *Chem. Rev.* **90** (1990) 383.

[97] S. M. Owen, Electron Counting in Clusters: a View of the Concepts. *Polyhedron* **7** (1988) 253.

[98] B.K. Teo, New Topological Electron-Counting Theory. *Inorg. Chem.* **23** (1984) 1251.

[99] W.B. Pearson, The Crystal Structures of Semiconductors and a General Valence Rule. *Acta Crystallogr.* **17** (1964) 1. Cf. also A. Keskhus and T. Rakke, *Structure and Bonding* **9** (1974) 45.

[100] R.J. Gillespie, Ring, Cage, and Cluster Compounds of the Main Group Elements. *J. Chem. Soc. Rev.* **1979**, 315.

[101] H. G. v. Schnering and W. Hönle, Bridging Chasms with Polyphosphides, *Chem. Rev.* **88** (1988) 243.

[102] H.G. v. Schnering, Homonuclear Bonds with Main Group Elements. *Angew. Chem. Intern. Ed.* **20** (1981) 33.

[103] B. Krebs, Thio and Seleno Compounds of Main Group Elements – New Inorganic Oligomers and Polymers. *Angew. Chem. Intern. Ed.* **22** (1983) 113.

[104] B. Krebs and G. Henkel, Transition Metal Thiolates — From Molecular Fragments of Sulfidic Solids to Models of Active Centers in Biomolecules. *Angew. Chem. Intern. Ed.* **30** (1991) 769.

[105] P. Böttcher, Tellurium-Rich Tellurides. *Angew. Chem. Intern. Ed.* **27** (1988) 759.

[106] H. Schäfer, B. Eisenmann and W. Müller, Zintl-Phases: Intermediate Forms between Metallic and Ionic Bonding. *Angew. Chem. Intern. Ed.* **12** (1973) 694.

[107] R. Nesper, Structure and Chemical Bonding in Zintl-Phases containing Lithium. *Progr. Solid State Chem.* **20** (1990) 1.

[108] R. Nesper, Chemical Bonds — Intermetallic Compounds. *Angew. Chem. Intern. Ed.* **30** (1991) 789.

[109] H. Schäfer, Semimetal Clustering in Intermetallic Phases. *J. Solid State Chem.* **57** (1985) 97.

[110] R. Chevrel, in: *Superconductor Materials Sciences Metallurgy, Fabrication and Applications* (S. Foner and B. B. Schwarz, eds.), Chap. 10. Plenum Press, New York 1981.

[111] J.D. Corbett, Polyatomic Zintl Anions of the Post-Transition Elements. *Chem. Rev.* **85** (1985) 383.

[112] A. Simon, Strukturchemie metallreicher Verbindungen, *Chemie in unserer Zeit* **10** (1976) 9.

[113] A. Simon, Condensed Metal Clusters. *Angew. Chem. Intern. Ed.* **20** (1981) 1.

[114] A. Simon, Clusters of Metals Poor in Valence Electrons – Structures, Bonding, Properties. *Angew. Chem. Intern. Ed.* **27** (1988) 159.

[115] J.D. Corbett, Extended Metal Metal Bonding in Halides of the Early Transition Metals. *Acc. Chem. Res.* **14** (1981) 239.

[116] T. Hughbanks, Bonding in Clusters and Condensed Culster Compounds that Extend in One, Two and Three Dimensions. *Prog. Solid State Chem.* **19** (1990) 329.

[117] R.L. Johnston and R. Hoffmann, Structure Bonding Relationships in the Laves Phases. *Z. Anorg. Allg. Chem.* **616** (1992) 105.

[118] T. Lundström, Preparation and Crystal Chemistry of some Refractory Borides and Phosphides, *Arkiv Kemi* **31** (1969) 227.

[119] P. Hagenmuller, Les Bronzes Oxygénés, *Progr. Solid State Chem.* **5** (1971) 71.

[120] M. Greenblatt, Molybdenum Oxide Bronzes with quasi Low-dimensional Properties. *Chem. Rev.* **88** (1988) 31.

[121] F. Hulliger, Crystal Chemistry of the Chalcogenides and Pnictides of the Transition Metals, *Structure and Bonding* **4** (1968) 82.

[122] S.C. Lee and R.H. Holm, Nonmolecular Metal Chalcogenide–Halide Solid State Compounds and their Molecular Analoga. *Angew. Chem. Intern. ed.* **29** (1990) 840.

[123] A. Kjeskhus and W.B. Pearson, Phases with the Nickel Arsenide and Closely-related Structures, *Progr. Solid State Chem.* **1** (1964) 83.

[124] D. Babel and A. Tressaud, Crystal Chemistry of Fluorides. In: *Inorganic Solid Fluorides* (P. Hagenmuller, ed.) Academic Press, Orlando–London 1985.

[125] M.T. Pope and A. Müller, Chemistry of Polyoxometallates: Variations of an Old Theme with Interdisciplinary Relations. *Angew. Chem. Intern. ed.* **30** (1991) 34.

[126] H. Müller-Buschbaum, Crystal Chemistry of Copper Oxometallates. *Angew. Chem. Intern. Ed.* **30** (1991) 723.

[127] M. Fliglharz, New Oxides in the $MoO_3–WO_3$ System. *Progr. Solid State Chem.* **19** (1989) 1.

[128] G. Meyer, The Syntheses and Structures of Complex Rare-earth Halides, *Progr. Solid State Chem.* **14** (1982) 141.

[129] J.V. Smith, Topochemistry of Zeolite and Related Materials. *Chem. Rev.* **88** (1988) 149.

[130] W. Bronger, Ternary Sulfides: A Model Case of the Relation between Crystal Structure and Magnetism. *Angew. Chem. Intern. Ed.* **20** (1981) 52.

[131] U. Müller, MX_4-Ketten aus kantenverknüpften Oktaedern: mögliche Kettenkonfigurationen und mögliche Kristallstrukturen. *Acta Crystallogr.* **B 37** (1981) 532.

[132] H. Bärnighausen, Group-Subgroup Relations between Space Groups: a Useful Tool in Crystal Chemistry. *MATCH, Communications in Mathematical Chemistry* **9** (1980) 139.

[133] G.O. Brunner, An Unconventional View of the Closest Sphere Packings. *Acta Crystallogr.* **A 27** (1971) 388.

[134] A. Meyer, Symmetriebeziehungen zwischen Kristallstrukturen des Formeltyps AX_2, ABX_4 und AB_2X_6 sowie deren Ordnungs- und Leerstellenvarianten. Dissertation, Univ. Karlsruhe (Germany) 1981.

Answers to the Problems

2.1 (a) 5.18; (b) 5.91; (c) 12.53.

2.2 (a) $Fe^oTi^oO_3^{[2n,2n]}$ or $Fe^{[6o]}Ti^{[6o]}O_3^{[2n,2n]}$; (b) $Cd^oCl_2^{[3n]}$; (c) $Mo^{[6p]}S_2^{[3n]}$; (d) $Cu_2^{[2l]}O^t$; (e) $Pt^{[4l]}S^{[4t]}$ or Pt^sS^t; (f) $Mg^{[16FK]}Cu_2^i$; (g) $Al_2^oMg_3^{do}Si_3^oO_{12}$; (h) $U^{[6p3c]}Cl_3^{[3n]}$.

2.3 CsCl, P; NaCl, F; sphalerite, F; CaF_2, F; rutile, P; CaC_2, I; NaN_3, P; ReO_3, P.

2.4 CsCl, 1; ZnS, 4; TiO_2, 2; $ThSi_2$, 4; ReO_3, 1; α-$ZnCl_2$, 4.

2.5 271.4 pm.

2.6 I(1)–I(2) 272.1 pm; I(2)–I(3) 250.0 pm; angle 178.4°.

2.7 210.2 and 213.2 pm; angle 101.8°.

2.8 W=O 177.5 pm; W$\cdots$O 216.0 pm; W–Br 244.4 pm; angle O=W–Br 97.2°; the coordination polyhedron is a distorted octahedron (cf. Fig. 135, p. 214).

2.9 Zr–O(1), 205.1, 216.3 and 205.7 pm; Zr–O(2), 218.9, 222.0, 228.5 and 215.1 pm; c.n. 7.

3.1 1st order (hysteresis observed).

3.2 It will melt at appprox. 0.1 GPa and refreeze at approx. 0.45 GPa, forming modification V.

3.3 It will freeze at appprox. 1.3 GPa forming modification VI.

3.4 $H_2O\cdot HF$ will crystallize, then, in addition, H_2O will freeze at $-72\,°C$.

3.5 Yes.

4.1 β-cristobalite could be converted to α and β-quartz.

4.2 At 1000 °C recrystallization will be faster.

4.3 BeF_2.

4.4 $-\frac{8}{1} + \frac{6\sqrt{3}}{2} + \frac{12\sqrt{3}}{2\sqrt{2}} - \frac{24\sqrt{3}}{\sqrt{10}}$.

4.5 (a) 687 kJ mol^{-1}; (b) 2965 kJ mol^{-1}; (c) 3517 kJ mol^{-1}.

5.1 F$\cdots$F in SiF_4 253 pm, van der Waals distance 294 pm; Cl$\cdots$Cl in $SiCl_4$ 330 pm, van der Waals distance 350 pm; I$\cdots$I in SiI_4 397 pm, van der Waals distance 396 pm; in SiF_4 and $SiCl_4$ the halogen atoms are squeezed.

5.2 WF_6 193, WCl_6 241, PCl_6^- 219, PBr_6^- 234, SbF_6^- 193, MnO_4^{2-} 166 pm; ReO_3 195, TiO_2 201, EuO 257, $CdCl_2$ 276 pm.

6.1 (a) rutile; (b) rutile; (c) neither (GeO_2 is actu-ally polymorphic, adopting the rutile and the quartz structures); (d) anti-CaF_2.

6.2 Mg^{2+} c.n. 8, Al^{3+} c.n. 6, Si^{4+} c.n. 4 (interchange of the c.n. of Mg^{2+} and Si^{4+} would also fulfil PAULING's rule, but c.n. 8 is rather improbable for Si^{4+}).

6.3 Since all cations have the same charge (+3), the electrostatic valence rule is of no help. The larger Y^{3+} ions will take the sites with c.n. 8.

6.4 No.

6.5 N is coordinated to Ag.

6.6 $s(Rb^+) = \frac{1}{10}$; $s(V^{4+}) = \frac{4}{5}$ $s(V^{5+}) = \frac{5}{4}$; $p_1 = 1.20$; $p_2 = 2.25$; $p_3 = 2.70$; $p_4 = 1.55$; $\bar{p}(V^{4+}) = 2.04$; $\bar{p}(V^{5+}) = 2.19$; Expected bond lengths: V^{4+}–O(1) 159 pm, V^{4+}–O(2) 197 pm, V^{5+}–O(2) 173 pm, V^{5+}–O(3) 180 pm, V^{5+}–O(4) 162 pm.

7.1 Linear: $BeCl_2$, Cl_3^-; angular: O_3^- (radical), S_3^{2-}; trigonal planar: BF_3; trigonal pyramidal PF_3, $TeCl_3^+$; T shaped: BrF_3, XeF_3^+; tetrahedral: $GeBr_4$, $AsCl_4^+$, $TiBr_4$, O_3BrF; square planar: ICl_4^-; trigonal bipyramidal with a missing equatorial vertex: SbF_4^-, $O_2ClF_2^-$ (F axial); trigonal bipyramid: $SbCl_5$, $SnCl_5^-$, O_2ClF_3 and O_3XeF_2 (O equatorial); square pyramidal: TeF_5^-; octahedral: $ClSF_5$.

7.2

Ta_2I_{10} like Nb_2Cl_{10} (cf. p. 54).

7.3 Trigonal bipyramid, CH_2 group in equatorial position perpendicular to equatorial plane. Derive it from an octahedron with bent S=C bonds.

7.4 (a) $SF_2 < SCl_2 < S_3^{2-} < S_3^- \approx OF_2$; (b) $H_3CNH_2 < [(H_3C)_2NH_2]^+$; (c) $PCl_2F_3 < PCl_3F_2$ $(=180°)$.

7.5 Bond lengths Al–Cl(terminal) < Al–Cl(bridge); angles Cl(bridge)–Al–Cl(bridge) $\approx 95°$ < Cl(bridge)–Al–Cl(terminal) $\approx 110°$ < Cl(terminal)–Al–Cl(terminal) $\approx 120°$.

7.6 $SnCl_3^-$; PF_6^-; $SnCl_6^{2-}$.

7.7 $BiBr_5^{2-}$ and TeI_6^{2-}.

8.1 $[Cr(OH_2)_6]^{2+}$, $[Mn(OH_2)_6]^{3+}$, $[Cu(NH_3)_6]^{2+}$.

8.2 $CrCl_4^-$ and $NiBr_4^{2-}$, elongated tetrahedron; $CuBr_4^{2-}$, flattened tetrahedron; $FeCl_4^{2-}$ could be slightly distorted.

8.3 Tetrahedral: $Co(CO)_4^-$, $Ni(PF_3)_4$, $Cu(OH)_4^{2-}$ (distorted); square: $PtCl_2(NH_3)_2$, $Pt(NH_3)_4^{2+}$, Au_2Cl_6.

8.4 (a) 2; (b) 1; (c) 2; (d) 2; (e) 1.

9.1 The band will broaden and the DOS will decrease.

9.2 It would look like the right part of Fig. 36.

9.3 The s band, the p_y band and the p_z band will shift to lower energy values at Γ and X', and to higher values at X and M; the p_x band will shift to higher values at Γ and X', and to lower values at X and M.

11.1 Shorter: BeO, BN; the same: BeS, BP, AlN; longer: AlP.

11.2 Longer bonds.

11.3 Under pressure AgI could adopt the NaCl structure (it actually does).

11.4 3.

11.5 Hg_2C should have the Cu_2O structure.

12.1 (a) simple ionic; (b) polyanionic; (c) polyanionic; (d) polyanionic; (e) polycationic; (f) polyanionic; (g) polycationic; (h) simple ionic.

12.2 (a), (b), (d).

12.3 (a)

$$\begin{array}{c} \overset{\ominus}{Te} \quad\quad \overset{\ominus}{Te} \\ {}_{\ominus}\diagdown{}^{Te}{}_{\diagup\ominus} \\ Al \diagup \diagdown Al \\ {}_{\ominus}\diagup {}^{Te}{}_{\diagdown\ominus} \\ Te \quad\quad Te \end{array}$$

(b)

$$\begin{array}{c} \overset{2\ominus}{Sb}\;\overset{2\ominus}{Sb}\;\overset{2\ominus}{Sb}\;\overset{2\ominus}{Sb} \\ \diagdown\diagup\;\;\diagdown\diagup \\ Sn \quad\quad Sn \\ \diagup\diagdown\;\;\diagup\diagdown \\ \underset{Sb}{{}_{\ominus}}\quad \underset{Sb}{{}_{\ominus}}\quad \underset{Sb}{{}_{\ominus}} \end{array}$$

(c) Layers as in elemental Sb;

(d)

$$\begin{array}{c} \overset{2\ominus}{Si} \\ {}^{2\ominus}\diagdown \;\diagup {}^{2\ominus} \\ Si \quad Si \end{array}$$

(e) $\overset{2\ominus}{P}{-}\overset{2\ominus}{P}$

12.4 (a) Wade ($26\ e^-$); (b) electron precise ($84\ e^-$); (c) $2e3c$ ($56\ e^-$); (d) electron precise ($72\ e^-$); (e) Wade ($86\ e^-$).

13.1 (a) $hhccc$ or 41; (b) $hhhc$ or 211.

13.2 (a) $ABACBC$; (b) $ABCACABCBCAB$; (c) $ABCBABACAB$.

14.1 (a) No (because of different structures); (b) yes; (c) no; (d) no; (e) no; (f) yes; (g) yes; (h) no.

14.2

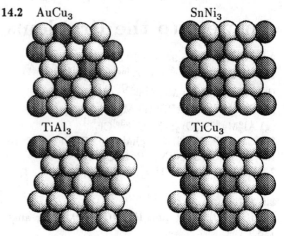

AuCu₃   SnNi₃

TiAl₃   TiCu₃

14.3 (a) CaF_2; (b) MgAgAs; (c) $MgCu_2Al$.

14.4 In both compounds each of the elements occupies two of the four different positions, but with different multiplicities: 3Cu:2Cu:2Zn:6Zn and 3Cu:2Al:2Al:6Cu.

14.5 Yes.

15.1 WO_3.

15.2 MX_4; this is the structure of a form of $ReCl_4$.

15.3 MX_4.

15.4 MoI_3 and TaS_3^{2-}.

15.5 Cristobalite.

16.1 MX_2.

16.2 Strands of face-sharing octahedra occur only in hexagonal closest-packing.

16.3 TiN, NaCl type; FeP, MnP type; FeSb, NiAs type; CoS, NiP type; CoSb, NiAs type.

16.4 In $CaBr_2$ and RhF_3 there is a three-dimensional linking of the octahedra; CdI_2 and BiI_3 consist of layers that can mutually be displaced.

16.5 $\frac{1}{7}$.

16.6 MgV_2O_4, normal; VMg_2O_4, inverse; $NiGa_2O_4$, inverse; $ZnCr_2S_4$, normal; $NiFe_2O_4$, inverse.

18.1 H_2O, $mm2$; $HCCl_3$, $3m$; BF_3, $\bar{6}2m$; XeF_4, $4/m\,2/m\,2/m$ (short $4/mmm$); $ClSF_5$, $4mm$; SF_6, $4/m\,\bar{3}\,2/m$ (short $m\bar{3}m$); cis-$SbF_4Cl_2^-$, $mm2$; $trans$-N_2F_2, $2/m$; $B(OH)_3$, $\bar{6}$; $Co(NO_2)_6^{3-}$, $2/m\bar{3}$ (short $m\bar{3}$).

18.2 Si_4^{6-}, $mm2$; As_4S_4, $\bar{4}2m$; P_4S_3, $3m$; Sn_5^{2-}, $\bar{6}2m$; Sn_9^{4-}, $4mm$; As_4^{6-}, 1; Sb_6^{8-}, 2; As_4^{4-}, $4/m\,2/m\,2/m$ (short $4/mmm$); P_6^{6-}, $\bar{3}2/m1$ (short $\bar{3}m$); P_7^{3-}, $3m$; P_{11}^{3-}, 3.

18.3 Tetrahedra, $2/m\,2/m\,2/m$ (short mmm) and $mm2$; octahedra, $4/m\,2/m\,2/m$ (short $4/mmm$), $\bar{8}2m$ and m.

18.4 Referred to the direction of the chain: translation vector, 2_1 axis and one mirror plane through each O atom (the mirror planes are perpendicular to the direction of reference); referred to the normal on the plane of the paper: mirror plane and one 2 axis through each Hg atom; referred to the direction vertical in the plane of the paper: glide plane and one 2 axis through each O atom; one inversion center in every Hg atom. If we define a coordinate system **a**,**b**,**c** with **a** perpendicular to the plane of the paper and **c** = translation vector, the Hermann–Mauguin symbol is $P(2/m\,2/c)2_1/m$; the parentheses designate the directions for which there is no translation symmetry.

18.5 hexagonal $M_x WO_3$, $P\,6/m\,2/m\,2/m$ (this is an idealized symmetry; actually the octahedra are slightly tilted and the real space group is $P6_322$); tetragonal $M_x WO_3$, $P\,4/m\,2_1/b\,2/m$ (short $P4/mbm$); NiAs, $P\,6_3/m\,2/m\,2/c$ (short $P6_3/mmc$); BaTiO$_3$, $P4mm$.

18.6 (a) $2/m\,2/m\,2/m$ (short mmm), orthorhombic; (b) $4/mmm$, tetragonal; (c) $\bar{3}2/m1$ (short $\bar{3}m$), trigonal; (d) $2/m$, monoclinic; (e) $6/m$, hexagonal; (f) 622, hexagonal; (g) 222, orthorhombic.

19.1 (a) klassengleich; (b) translationengleich; (c) klassengleich; (d) isomorphic; (e) translationengleich; (f) translationengleich; (g) translationengleich.

19.2 β-Sn will form triplets.

19.3 Yes.

Index